MW01195619

Incompleteness and Computability

The Open Logic Project

Instigator

Richard Zach, *University of Calgary*

Editorial Board

Aldo Antonelli,[†] *University of California, Davis*
Andrew Arana, *Université Paris I Panthénon–Sorbonne*
Jeremy Avigad, *Carnegie Mellon University*
Tim Button, *University College London*
Walter Dean, *University of Warwick*
Gillian Russell, *University of North Carolina*
Nicole Wyatt, *University of Calgary*
Audrey Yap, *University of Victoria*

Contributors

Samara Burns, *University of Calgary*
Dana Hägg, *University of Calgary*
Zesen Qian, *Carnegie Mellon University*

Incompleteness and Computability

An Open Introduction to Gödel's Theorems

Remixed by Richard Zach

FALL 2019

The Open Logic Project would like to acknowledge the generous support of the Taylor Institute of Teaching and Learning of the University of Calgary, and the Alberta Open Educational Resources (ABOER) Initiative, which is made possible through an investment from the Alberta government.

Cover illustrations by Matthew Leadbeater, used under a Creative Commons Attribution-NonCommercial 4.0 International License.

Typeset in Baskervald X and Nimbus Sans by LaTeX.

This version of *Incompleteness and Computability* is revision 502219d (2019-11-09), with content generated from *Open Logic Text* revision 1cdcec1 (2019-11-09). Free download at:
`https://ic.openlogicproject.org/`

Incompleteness and Computability by Richard Zach is licensed under a Creative Commons Attribution 4.0 International License. It is based on *The Open Logic Text* by the Open Logic Project, used under a Creative Commons Attribution 4.0 International License.

Contents

About this Book

This is a textbook on Gödel's incompleteness theorems and recursive function theory. I use it as the main text when I teach Philosophy 479 (Logic III) at the University of Calgary. It is based on material from the Open Logic Project.

As its name suggests, the course is the third in a sequence, so students (and hence readers of this book) are expected to be familiar with first-order logic already. (Logic I uses the text *forall x: Calgary*, and Logic II another textbook based on the OLP, *Sets, Logic, Computation*.) The material assumed from Logic II, however, is included as appendices B and C.

Logic III is a thirteen-week course, meeting three hours per week. This is typically enough to cover the material in chapters 1 to 5 and either chapter 6 or chapter 8, depending on student interest. You may want to spend more time on the basics of first-order logic and especially on natural deduction, if students are not already familiar with it. Note that when provability in arithmetical theories (such as **Q** and **PA**) is discussed in the main text, the proofs of provability claims are not given using a specific proof system. Rather, that certain claims follow from the axioms by first-order logic is justified intuitively. However, appendix A contains a number of examples of actual natural deduction derivations from the axioms of **Q**.

Acknowledgments

The material in the OLP used in chapters 1 to 5 and 8 was based originally on Jeremy Avigad's lecture notes on "Computability and Incompleteness," which he contributed to the OLP. I have heavily revised and expanded this material. The lecture notes, e.g., based theories of arithmetic on an axiomatic proof system. Here, we use Gentzen's standard natural deduction system (described in appendix C), which requires dealing with trees primitive recursively (in section 2.12) and a more complicated approach to the arithmetization of derivations (in section 3.6). The material in chapter 8 was also expanded by Zesen Qian during his stay in Calgary as a Mitacs summer intern.

The material in the OLP on model theory and models of arithmetic in chapter 6 was originally taken from Aldo Antonelli's lecture notes on "The Completeness of Classical Propositional and Predicate Logic," which he contributed to the OLP before his untimely death in 2015.

The biographies of logicians in appendix D and much of the material in appendix C are originally due to Samara Burns. Dana Hägg originally worked on the material in appendix B.

CHAPTER 1

Introduction to Incompleteness

1.1 Historical Background

In this section, we will briefly discuss historical developments that will help put the incompleteness theorems in context. In particular, we will give a very sketchy overview of the history of mathematical logic; and then say a few words about the history of the foundations of mathematics.

The phrase "mathematical logic" is ambiguous. One can interpret the word "mathematical" as describing the subject matter, as in, "the logic of mathematics," denoting the principles of mathematical reasoning; or as describing the methods, as in "the mathematics of logic," denoting a mathematical study of the principles of reasoning. The account that follows involves mathematical logic in both senses, often at the same time.

The study of logic began, essentially, with Aristotle, who lived approximately 384–322 BCE. His *Categories*, *Prior analytics*, and *Posterior analytics* include systematic studies of the principles of scientific reasoning, including a thorough and systematic study of the syllogism.

Aristotle's logic dominated scholastic philosophy through the middle ages; indeed, as late as eighteenth century Kant maintained that Aristotle's logic was perfect and in no need of revision. But the theory of the syllogism is far too limited to model anything but the most superficial aspects of mathematical reasoning. A century earlier, Leibniz, a contemporary of Newton's, imagined a complete "calculus" for logical reasoning, and made some rudimentary steps towards designing such a calculus, essentially describing a version of propositional logic.

The nineteenth century was a watershed for logic. In 1854 George Boole wrote *The Laws of Thought*, with a thorough algebraic study of propositional logic that is not far from modern presentations. In 1879 Gottlob Frege published his *Begriffsschrift* (Concept writing) which extends propositional logic with quantifiers and relations, and thus includes first-order logic. In fact, Frege's logical systems included higher-order logic as well, and more. In his *Basic Laws of Arithmetic*, Frege set out to show that all of arithmetic could be derived in his Begriffsschrift from purely logical assumption. Unfortunately, these assumptions turned out to be inconsistent, as Russell showed in 1902. But setting aside the inconsistent axiom, Frege more or less invented modern logic singlehandedly, a startling achievement. Quantificational logic was also developed independently by algebraically-minded thinkers after Boole, including Peirce and Schröder.

Let us now turn to developments in the foundations of mathematics. Of course, since logic plays an important role in mathematics, there is a good deal of interaction with the developments just described. For example, Frege developed his logic with the explicit purpose of showing that all of mathematics could be based solely on his logical framework; in particular, he wished to show that mathematics consists of a priori *analytic* truths instead of, as Kant had maintained, a priori *synthetic* ones.

Many take the birth of mathematics proper to have occurred with the Greeks. Euclid's *Elements*, written around 300 B.C., is already a mature representative of Greek mathematics, with its emphasis on rigor and precision. The definitions and proofs in Euclid's *Elements* survive more or less in tact in high school geometry textbooks today (to the extent that geometry is still taught in high schools). This model of mathematical reasoning has been held to be a paradigm for rigorous argumentation not only in mathematics but in branches of philosophy as well. (Spinoza even presented moral and religious arguments in the Euclidean style, which is strange to see!)

Calculus was invented by Newton and Leibniz in the seventeenth century. (A fierce priority dispute raged for centuries, but most scholars today hold that the two developments were for the most part independent.) Calculus involves reasoning about, for example, infinite sums of infinitely small quantities; these features fueled criticism by Bishop Berkeley, who argued that belief in God was no less rational than the mathematics of his time. The methods of calculus were widely used in the eighteenth century, for example by Leonhard Euler, who used calculations involving infinite sums with dramatic results.

In the nineteenth century, mathematicians tried to address Berkeley's criticisms by putting calculus on a firmer foundation. Efforts by Cauchy, Weierstrass, Bolzano, and others led to our contemporary definitions of limits, continuity, differentiation, and integration in terms of "epsilons and deltas," in other words, devoid of any reference to infinitesimals. Later in the century, mathematicians tried to push further, and explain all aspects of calculus, including the real numbers themselves, in terms of the natural numbers. (Kronecker: "God created the whole numbers, all else is the work of man.") In 1872, Dedekind wrote "Continuity and the irrational numbers," where he showed how to "construct" the real numbers as sets of rational numbers (which, as you know, can be viewed as pairs of natural numbers); in 1888 he wrote "Was sind und was sollen die Zahlen" (roughly, "What are the natural numbers, and what should they be?") which aimed to explain the natural numbers in purely "logical" terms. In 1887 Kronecker wrote "Über den Zahlbegriff" ("On the concept of number") where he spoke of representing all mathematical object in terms of the integers; in 1889 Giuseppe Peano gave formal, symbolic axioms for the natural numbers.

The end of the nineteenth century also brought a new boldness in dealing with the infinite. Before then, infinitary objects and structures (like the set of natural numbers) were treated gingerly; "infinitely many" was understood as "as many as you want," and "approaches in the limit" was understood as "gets as close as you want." But Georg Cantor showed that it was possible to take the infinite at face value. Work by Cantor, Dedekind, and others help to introduce the general set-theoretic understanding of mathematics that is now widely accepted.

This brings us to twentieth century developments in logic and foundations. In 1902 Russell discovered the paradox in Frege's logical system. In 1904 Zermelo proved Cantor's well-ordering principle, using the so-called "axiom of choice"; the legitimacy of this axiom prompted a good deal of debate. Between 1910 and 1913 the three volumes of Russell and Whitehead's *Principia Mathematica* appeared, extending the Fregean program of establishing mathematics on logical grounds. Unfortunately, Russell and Whitehead were forced to adopt two principles that seemed hard to justify as purely logical: an axiom of infinity and an axiom of "reducibility." In the 1900's Poincaré criticized the use of "impredicative definitions" in mathematics, and in the 1910's Brouwer began proposing to refound all of mathematics in an "intuitionistic" basis, which avoided the use of the law of the excluded middle ($A \lor \neg A$).

Strange days indeed! The program of reducing all of mathematics to logic is now referred to as "logicism," and is commonly viewed as having failed, due to the difficulties mentioned above. The program of developing mathematics in terms of intuitionistic mental constructions is called "intuitionism," and is viewed as posing overly severe restrictions on everyday mathematics. Around the turn of the century, David Hilbert, one of the most influential mathematicians of all time, was a strong supporter of the new, abstract methods introduced by Cantor and Dedekind: "no one will drive us from the paradise that Cantor has created for us." At the same time, he was sensitive to foundational criticisms of these new methods (oddly enough, now called "classical"). He proposed a way of having one's cake and eating it too:

1. Represent classical methods with formal axioms and rules; represent mathematical questions as formulas in an axiomatic system.

2. Use safe, "finitary" methods to prove that these formal deductive systems are consistent.

Hilbert's work went a long way toward accomplishing the first goal. In 1899, he had done this for geometry in his celebrated book *Foundations of geometry*. In subsequent years, he and a number of his students and collaborators worked on other areas of mathematics to do what Hilbert had done for geometry. Hilbert himself gave axiom systems for arithmetic and analysis. Zermelo gave an axiomatization of set theory, which was expanded on by Fraenkel, Skolem, von Neumann, and others. By the mid-1920s, there were two approaches that laid claim to the title of an axiomatization of "all" of mathematics, the *Principia mathematica* of Russell and Whitehead, and what came to be known as Zermelo-Fraenkel set theory.

In 1921, Hilbert set out on a research project to establish the goal of proving these systems to be consistent. He was aided in this project by several of his students, in particular Bernays, Ackermann, and later Gentzen. The basic idea for accomplishing this goal was to cast the question of the possibility of a derivation of an inconsistency in mathmatics as a combinatorial problem about possible sequences of symbols, namely possible sequences of sentences which meet the criterion of being a correct derivation of, say, $A \land \neg A$ from the axioms of an axiom system for arithmetic, analysis, or set theory. A proof of the impossibility of such a sequence of symbols would—since it is itself a mathematical proof—be formalizable in these axiomatic systems. In other words, there would be some sentence Con which states that, say, arithmetic is consistent. Moreover, this sentence should be provable in the systems in question, especially if its proof requires only very restricted, "finitary" means.

The second aim, that the axiom systems developed would settle every mathematical question, can be made precise in two ways. In one way, we can formulate it as follows: For any sentence A in the language of an axiom system for mathematics, either A or $\neg A$ is provable from the axioms. If this were true, then there would be no sentences which can neither be proved nor refuted on the basis of the axioms, no questions which the axioms do not settle. An axiom system with this property is called *complete.* Of course, for any given sentence it might still be a difficult task to determine which of the two alternatives holds. But in principle there should be a method to do so. In fact, for the axiom and derivation systems considered by Hilbert, completeness would imply that such a method exists—although Hilbert did not realize this. The second way to interpret the question would be this stronger requirement: that there be a mechanical, computational method which would determine, for a given sentence A, whether it is derivable from the axioms or not.

In 1931, Gödel proved the two "incompleteness theorems," which showed that this program could not succeed. There is no axiom system for mathematics which is complete, specifically, the sentence that expresses the consistency of the axioms is a sentence which can neither be proved nor refuted.

This struck a lethal blow to Hilbert's original program. However, as is so often the case in mathematics, it also opened up exciting new avenues for research. If there is no one, all-encompassing formal system of mathematics, it makes sense to develop more circumscribesd systems and investigate what can be proved in them. It also makes sense to develop less restricted methods of proof for establishing the consistency of these systems, and to find ways to measure how hard it is to prove their consistency. Since Gödel showed that (almost) every formal system has questions it cannot settle, it makes sense to look for "interesting" questions a given formal system cannot settle, and to figure out how strong a formal system has to be to settle them. To the present day, logicians have been pursuing these questions in a new mathematical discipline, the theory of proofs.

1.2 Definitions

In order to carry out Hilbert's project of formalizing mathematics and showing that such a formalization is consistent and complete, the first order of business would be that of picking a language, logical framework, and a system of axioms. For our purposes, let us suppose that mathematics can be formalized in a first-order language, i.e., that there is some set of constant symbols, function symbols, and predicate symbols which, together with the connectives and quatifiers of first-order logic, allow us to express the claims of mathematics. Most people agree that such a language exists: the language of set theory, in which $\in$ is the only non-logical symbol. That such a simple language is so expressive is of course a very implausible claim at first sight, and it took a lot of work to establish that practically of all mathematics can be expressed in this very austere vocabulary. To keep things simple, for now, let's restrict our discussion to arithmetic, so the part of mathematics that just deals with the natural numbers $\mathbb{N}$. The natural language in which to express facts of arithmetic is $\mathcal{L}_A$. $\mathcal{L}_A$ contains a single two-place predicate symbol $<$, a single constant symbol 0, one one-place function symbol $\prime$, and two two-place function symbols $+$ and $\times$.

Definition 1.1. A set of sentences Γ is a *theory* if it is closed under entailment, i.e., if $\Gamma = \{A : \Gamma \vDash A\}$.

There are two easy ways to specify theories. One is as the set of sentences true in some structure. For instance, consider the structure for $\mathcal{L}_A$ in which the domain is $\mathbb{N}$ and all non-logical symbols are interpreted as you would expect.

Definition 1.2. The *standard model of arithmetic* is the structure $\mathfrak{N}$ defined as follows:

1. $|\mathfrak{N}| = \mathbb{N}$

2. $\mathrm{o}^{\mathfrak{N}} = 0$

3. $\prime^{\mathfrak{N}}(n) = n + 1$ for all $n \in \mathbb{N}$

4. $+^{\mathfrak{N}}(n, m) = n + m$ for all $n, m \in \mathbb{N}$

5. $\times^{\mathfrak{N}}(n, m) = n \cdot m$ for all $n, m \in \mathbb{N}$

6. $<^{\mathfrak{N}} = \{\langle n, m \rangle : n \in \mathbb{N}, m \in \mathbb{N}, n < m\}$

Note the difference between $\times$ and $\cdot$: $\times$ is a symbol in the language of arithmetic. Of course, we've chosen it to remind us of multiplication, but $\times$ is not the multiplication operation but a two-place function symbol (officially, f_1^2. By contrast, $\cdot$ *is* the ordinary multiplication function. When you see something like $n \cdot m$, we mean the product of the numbers n and m; when you see something like $x \times y$ we are talking about a term in the language of arithmetic. In the standard model, the function symbol times is interpreted as the function $\cdot$ on the natural numbers. For addition, we use $+$ as both the function symbol of the language of arithmetic, and the addition function on the natural numbers. Here you have to use the context to determine what is meant.

Definition 1.3. The theory of *true arithmetic* is the set of sentences satisfied in the standard model of arithmetic, i.e.,

$$\mathbf{TA} = \{A : \mathfrak{N} \vDash A\}.$$

TA is a theory, for whenever $\mathbf{TA} \vDash A$, A is satisfied in every structure which satisfies **TA**. Since $\mathfrak{M} \vDash \mathbf{TA}$, $\mathfrak{M} \vDash A$, and so $A \in \mathbf{TA}$.

The other way to specify a theory Γ is as the set of sentences entailed by some set of sentences Γ_0. In that case, Γ is the "closure" of Γ_0 under entailment. Specifying a theory this way is only interesting if Γ_0 is explicitly specified, e.g., if the elements of Γ_0 are listed. At the very least, Γ_0 has to be decidable, i.e., there has to be a computable test for when a sentence counts as an element of Γ_0 or not. We call the sentences in Γ_0 *axioms* for Γ, and Γ *axiomatized* by Γ_0.

Definition 1.4. A theory Γ is *axiomatized* by Γ_0 iff

$$\Gamma = \{A : \Gamma_0 \vDash A\}$$

Definition 1.5. The theory **Q** axiomatized by the following sentences is known as "Robinson's **Q**" and is a very simple theory of arithmetic.

$$\forall x\, \forall y\, (x' = y' \to x = y) \tag{Q_1}$$
$$\forall x\, 0 \neq x' \tag{Q_2}$$
$$\forall x\, (x = 0 \lor \exists y\, x = y') \tag{Q_3}$$
$$\forall x\, (x + 0) = x \tag{Q_4}$$
$$\forall x\, \forall y\, (x + y') = (x + y)' \tag{Q_5}$$
$$\forall x\, (x \times 0) = 0 \tag{Q_6}$$
$$\forall x\, \forall y\, (x \times y') = ((x \times y) + x) \tag{Q_7}$$
$$\forall x\, \forall y\, (x < y \leftrightarrow \exists z\, (z' + x) = y) \tag{Q_8}$$

The set of sentences $\{Q_1, \dots, Q_8\}$ are the axioms of **Q**, so **Q** consists of all sentences entailed by them:

$$\mathbf{Q} = \{A : \{Q_1, \dots, Q_8\} \vDash A\}.$$

Definition 1.6. Suppose $A(x)$ is a formula in $\mathcal{L}_A$ with free variables x and $y_1, \dots, y_n$. Then any sentence of the form

$$\forall y_1 \dots \forall y_n\, ((A(0) \land \forall x\, (A(x) \to A(x'))) \to \forall x\, A(x))$$

is an instance of the *induction schema*.

Peano arithmetic **PA** is the theory axiomatized by the axioms of **Q** together with all instances of the induction schema.

Every instance of the induction schema is true in $\mathbf{N}$. This is easiest to see if the formula A only has one free variable x. Then $A(x)$ defines a subset X_A of $\mathbb{N}$ in $\mathbf{N}$. X_A is the set of all $n \in \mathbb{N}$ such that $\mathbf{N}, s \vDash A(x)$ when $s(x) = n$. The corresponding instance of the induction schema is

$$((A(0) \land \forall x\, (A(x) \rightarrow A(x'))) \rightarrow \forall x\, A(x)).$$

If its antecedent is true in $\mathbf{N}$, then $0 \in X_A$ and, whenever $n \in X_A$, so is $n + 1$. Since $0 \in X_A$, we get $1 \in X_A$. With $1 \in X_A$ we get $2 \in X_A$. And so on. So for every $n \in \mathbb{N}$, $n \in X_A$. But this means that $\forall x\, A(x)$ is satisfied in $\mathbf{N}$.

Both **Q** and **PA** are axiomatized theories. The big question is, how strong are they? For instance, can **PA** prove all the truths about $\mathbb{N}$ that can be expressed in $\mathcal{L}_A$? Specifically, do the axioms of **PA** settle all the questions that can be formulated in $\mathcal{L}_A$?

Another way to put this is to ask: Is **PA** = **TA**? **TA** obviously does prove (i.e., it includes) all the truths about $\mathbb{N}$, and it settles all the questions that can be formulated in $\mathcal{L}_A$, since if A is a sentence in $\mathcal{L}_A$, then either $\mathbf{N} \vDash A$ or $\mathbf{N} \vDash \neg A$, and so either $\mathbf{TA} \vDash A$ or $\mathbf{TA} \vDash \neg A$. Call such a theory *complete.*

Definition 1.7. A theory Γ is *complete* iff for every sentence A in its language, either $\Gamma \vDash A$ or $\Gamma \vDash \neg A$.

By the Completeness Theorem, $\Gamma \vDash A$ iff $\Gamma \vdash A$, so Γ is complete iff for every sentence A in its language, either $\Gamma \vdash A$ or $\Gamma \vdash \neg A$.

Another question we are led to ask is this: Is there a computational procedure we can use to test if a sentence is in **TA**, in **PA**, or even just in **Q**? We can make this more precise by defining when a set (e.g., a set of sentences) is decidable.

Definition 1.8. A set X is *decidable* iff there is a computational procedure which on input x returns 1 if $x \in X$ and 0 otherwise.

So our question becomes: Is **TA** (**PA**, **Q**) decidable?

The answer to all these questions will be: no. None of these theories are decidable. However, this phenomenon is not specific to these particular theories. In fact, *any* theory that satisfies certain conditions is subject to the same results. One of these conditions, which **Q** and **PA** satisfy, is that they are axiomatized by a decidable set of axioms.

Definition 1.9. A theory is *axiomatizable* if it is axiomatized by a decidable set of axioms.

Example 1.10. Any theory axiomatized by a finite set of sentences is axiomatizable, since any finite set is decidable. Thus, **Q**, for instance, is axiomatizable.

Schematically axiomatized theories like **PA** are also axiomatizable. For to test if B is among the axioms of **PA**, i.e., to compute the function χ_X where $\chi_X(B) = 1$ if B is an axiom of **PA** and $= 0$ otherwise, we can do the following: First, check if B is one of the axioms of **Q**. If it is, the answer is "yes" and the value of $\chi_X(B) = 1$. If not, test if it is an instance of the induction schema. This can be done systematically; in this case, perhaps it's easiest to see that it can be done as follows: Any instance of the induction schema begins with a number of universal quantifiers, and then a sub-formula that is a conditional. The consequent of that conditional is $\forall x\, A(x, y_1, \dots, y_n)$ where x and $y_1, \dots, y_n$ are all the free variables of A and the initial quantifiers of B bind the variables $y_1, \dots, y_n$. Once we have extracted this A and checked that its free variables match the variables bound by the universal qauntifiers at the front and $\forall x$, we go on to check that the antecedent of the conditional matches

$$A(0, y_1, \dots, y_n) \land \forall x\,(A(x, y_1, \dots, y_n) \to A(x', y_1, \dots, y_n))$$

Again, if it does, B is an instance of the induction schema, and if it doesn't, B isn't.

In answering this question—and the more general question of which theories are complete or decidable—it will be useful to consider also the following definition. Recall that a set X is countable iff it is empty or if there is a surjective function $f : \mathbb{N} \to X$. Such a function is called an enumeration of X.

Definition 1.11. A set X is called *computably enumerable* (c.e. for short) iff it is empty or it has a computable enumeration.

In addition to axiomatizability, another condition on theories to which the incompleteness theorems apply will be that they are strong enough to prove basic facts about computable functions and decidable relations. By "basic facts," we mean sentences which express what the values of computable functions are for each of their arguments. And by "strong enough" we mean that the theories in question count these sentences among its theorems. For instance, consider a prototypical computable function: addition. The value of $+$ for arguments 2 and 3 is 5, i.e., $2+3 = 5$. A sentence in the language of arithmetic that expresses that the value of $+$ for arguments 2 and 3 is 5 is: $(\overline{2} + \overline{3}) = \overline{5}$. And, e.g., **Q** proves this sentence. More generally, we would like there to be, for each computable function $f(x_1, x_2)$ a formula $A_f(x_1, x_2, y)$ in $\mathcal{L}_A$ such that $\mathbf{Q} \vdash A_f(\overline{n_1}, \overline{n_2}, \overline{m})$ whenever $f(n_1, n_2) = m$. In this way, **Q** proves that the value of f for arguments n_1, n_2 is m. In fact, we require that it proves a bit more, namely that no other number is the value of f for arguments n_1, n_2. And the same goes for decidable relations. This is made precise in the following two definitions.

Definition 1.12. A formula $A(x_1, \dots, x_k, y)$ *represents* the function $f\colon \mathbb{N}^k \to \mathbb{N}$ in Γ iff whenever $f(n_1, \dots, n_k) = m$, then

1. $\Gamma \vdash A(\overline{n_1}, \dots, \overline{n_k}, \overline{m})$, and
2. $\Gamma \vdash \forall y(A(\overline{n_1}, \dots, \overline{n_k}, y) \to y = \overline{m})$.

Definition 1.13. A formula $A(x_1, \dots, x_k)$ *represents* the relation $R \subseteq \mathbb{N}^k$ iff,

1. whenever $R(n_1, \dots, n_k)$, $\Gamma \vdash A(\overline{n_1}, \dots, \overline{n_k})$, and
2. whenever not $R(n_1, \dots, n_k)$, $\Gamma \vdash \neg A(\overline{n_1}, \dots, \overline{n_k})$.

A theory is "strong enough" for the incompleteness theorems to apply if it represents all computable functions and all decidable relations. **Q** and its extensions satisfy this condition, but it will take us a while to establish this—it's a non-trivial fact about the kinds of things **Q** can prove, and it's hard to show because **Q** has only a few axioms from which we'll have to prove all these facts. However, **Q** is a very weak theory. So although it's hard to prove that **Q** represents all computable functions, most interesting theories are stronger than **Q**, i.e., prove more than **Q** does. And if **Q** proves something, any stronger theory does; since **Q** represents all computable functions, every stronger theory does. This means that many interesting theories meet this condition of the incompleteness theorems. So our hard work will pay off, since it shows that the incompletess theorems apply to a wide range of theories. Certainly, any theory aiming to formalize "all of mathematics" must prove everything that **Q** proves, since it should at the very least be able to capture the results of elementary computations. So any theory that is a candidate for a theory of "all of mathematics" will be one to which the incompleteness theorems apply.

1.3 Overview of Incompleteness Results

Hilbert expected that mathematics could be formalized in an axiomatizable theory which it would be possible to prove complete and decidable. Moreover, he aimed to prove the consistency of this theory with very weak, "finitary," means, which would defend classical mathematics agianst the challenges of intuitionism. Gödel's incompleteness theorems showed that these goals cannot be achieved.

Gödel's first incompleteness theorem showed that a version of Russell and Whitehead's *Principia Mathematica* is not complete. But the proof was actually very general and applies to a wide variety of theories. This means that it wasn't just that *Principia Mathematica* did not manage to completely capture mathematics, but that *no* acceptable theory does. It took a while to isolate the features of theories that suffice for the incompleteness theorems to apply, and to generalize Gödel's proof to apply make it depend only on these features. But we are now in a position to state a very general version of the first incompleteness theorem for theories in the language $\mathcal{L}_A$ of arithmetic.

Theorem 1.14. *If Γ is a consistent and axiomatizable theory in $\mathcal{L}_A$ which represents all computable functions and decidable relations, then Γ is not complete.*

To say that Γ is not complete is to say that for at least one sentence A, $\Gamma \nvdash A$ and $\Gamma \nvdash \neg A$. Such a sentence is called *independent* (of Γ). We can in fact relatively quickly prove that there must be independent sentences. But the power of Gödel's proof of the theorem lies in the fact that it exhibits a *specific example* of such an independent sentence. The intriguing construction produces a sentence G_Γ, called a *Gödel sentence* for Γ, which is unprovable because in Γ, G_Γ is equivalent to the claim that G_Γ is unprovable in Γ. It does so *constructively*, i.e., given an axiomatization of Γ and a description of the proof system, the proof gives a method for actually writing down G_Γ.

The construction in Gödel's proof requires that we find a way to express in $\mathcal{L}_A$ the properties of and operations on terms and formulas of $\mathcal{L}_A$ itself. These include properties such as "A is a sentence," "δ is a derivation of A," and operations such as $A[t/x]$. This way must (a) express these properties and relations via a "coding" of symbols and sequences thereof (which is what terms, formulas, derivations, etc. are) as natural numbers (which is what $\mathcal{L}_A$ can talk about). It must (b) do this in such a way that Γ will prove the relevant facts, so we must show that these properties are coded by decidable properties of natural numbers and the operations correspond to computable functions on natural numbers. This is called "arithmetization of syntax."

Before we investigate how syntax can be arithmetized, however, we will consider the condition that Γ is "strong enough," i.e., represents all computable functions and decidable relations. This requires that we give a precise definition of "computable." This can be done in a number of ways, e.g., via the model of Turing machines, or as those functions computable by programs in some general-purpose programming language. Since our aim is to represent these functions and relations in a theory in the language $\mathcal{L}_A$, however, it is best to pick a simple definition of computability of just numerical functions. This is the notion of *recursive function*. So we will first discuss the recursive functions. We will then show that **Q** already represents all recursive functions and relations. This will allow us to apply the incompleteness theorem to specific theories such as **Q** and **PA**, since we will have established that these are examples of theories that are "strong enough."

The end result of the arithmetization of syntax is a formula $\mathrm{Prov}_\Gamma(x)$ which, via the coding of formulas as numbers, expresses provability from the axioms of Γ. Specifically, if A is coded by the number n, and $\Gamma \vdash A$, then $\Gamma \vdash \mathrm{Prov}_\Gamma(\overline{n})$. This "provability predicate" for Γ allows us also to express, in a certain sense, the consistency of Γ as a sentence of $\mathcal{L}_A$: let the "consistency statement" for Γ be the sentence $\neg\mathrm{Prov}_\Gamma(\overline{n})$, where we take n to be the code of a contradiction, e.g., of $\bot$. The second incompleteness theorem states that consistent axiomatizable theories also do not prove their own consistency statements. The conditions required for this theorem to apply are a bit more stringent than just that the theory represents all computable functions and decidable relations, but we will show that **PA** satisifes them.

1.4 Undecidability and Incompleteness

Gödel's proof of the incompleteness theorems require arithmetization of syntax. But even without that we can obtain some nice results just on the assumtion that a theory represents all decidable relations. The proof is a diagonal argument similar to the proof of the undecidability of the halting problem.

Theorem 1.15. *If Γ is a consistent theory that represents every decidable relation, then Γ is not decidable.*

Proof. Suppose Γ were decidable. We show that if Γ represents every decidable relation, it must be inconsistent.

Decidable properties (one-place relations) are represented by formulas with one free variable. Let $A_0(x)$, $A_1(x)$, ..., be a computable enumeration of all such formulas. Now consider the following set $D \subseteq \mathbb{N}$:

$$D = \{n : \Gamma \vdash \neg A_n(\overline{n})\}$$

The set D is decidable, since we can test if $n \in D$ by first computing $A_n(x)$, and from this $\neg A_n(\overline{n})$. Obviously, substituting the term $\overline{n}$ for every free occurrence of x in $A_n(x)$ and prefixing $A(\overline{n})$ by $\neg$ is a mechanical matter. By assumption, Γ is decidable, so we can test if $\neg A(\overline{n}) \in \Gamma$. If it is, $n \in D$, and if it isn't, $n \notin D$. So D is likewise decidable.

Since Γ represents all decidable properties, it represents D. And the formulas which represent D in Γ are all among $A_0(x)$, $A_1(x)$, ... So let d be a number such that $A_d(x)$ represents D in Γ. If $d \notin D$, then, since $A_d(x)$ represents D, $\Gamma \vdash \neg A_d(\overline{d})$.

But that means that d meets the defining condition of D, and so $d \in D$. This contradicts $d \notin D$. So by indirect proof, $d \in D$.

Since $d \in D$, by the definition of D, $\Gamma \vdash \neg A_d(\overline{d})$. On the other hand, since $A_d(x)$ represents D in Γ, $\Gamma \vdash A_d(\overline{d})$. Hence, Γ is inconsistent. □

The preceding theorem shows that no theory that represents all decidable relations can be decidable. We will show that **Q** does represent all decidable relations; this means that all theories that include **Q**, such as **PA** and **TA**, also do, and hence also are not decidable.

We can also use this result to obtain a weak version of the first incompleteness theorem. Any theory that is axiomatizable and complete is decidable. Consistent theories that are axiomatizable and represent all decidable properties then cannot be complete.

Theorem 1.16. *If Γ is axiomatizable and complete it is decidable.*

Proof. Any inconsistent theory is decidable, since inconsistent theories contain all sentences, so the answer to the question "is $A \in \Gamma$" is always "yes," i.e., can be decided.

So suppose Γ is consistent, and furthermore is axiomatizable, and complete. Since Γ is axiomatizable, it is computably enumerable. For we can enumerate all the correct derivations from the axioms of Γ by a computable function. From a correct derivation we can compute the sentence it derives, and so together there is a computable function that enumerates all theorems of Γ. A sentence is a theorem of Γ iff $\neg A$ is not a theorem, since Γ is consistent and complete. We can therefore decide if $A \in \Gamma$ as follows. Enumerate all theorems of Γ. When A appears on this list, we know that $\Gamma \vdash A$. When $\neg A$ appears on this list, we know that $\Gamma \nvdash A$. Since Γ is complete, one of these cases eventually obtains, so the procedure eventually produces and answer. □

Corollary 1.17. *If Γ is consistent, axiomatizable, and represents every decidable property, it is not complete.*

Proof. If Γ were complete, it would be decidable by the previous theorem (since it is axiomatizable and consistent). But since Γ represents every decidable property, it is not decidable, by the first theorem. □

Once we have established that, e.g., **Q**, represents all decidable properties, the corollary tells us that **Q** must be incomplete. However, its proof does not provide an example of an independent sentence; it merely shows that such a sentence must exist. For this, we have to arithmetize syntax and follow Gödel's original proof idea. And of course, we still have to show the first claim, namely that **Q** does, in fact, represent all decidable properties.

It should be noted that not every *interesting* theory is incomplete or undecidable. There are many theories that are sufficiently strong to describe interesting mathematical facts that do not satisfy the conditions of Gödel's result. For instance, **Pres** $= \{A \in \mathcal{L}_{A^+} : N \vDash A\}$, the set of sentences of the language of arithmetic without $\times$ true in the standard model, is both complete and decidable. This theory is called Presburger arithmetic, and proves all the truths about natural numbers that can be formulated just with 0, $'$, and $+$.

Summary

Hilbert's program aimed to show that all of mathematics could be formalized in an axiomatized theory in a formal language, such as the language of arithmetic or of set theory. He believed that such a theory would be **complete**. That is, for every sentence A, either $\mathbf{T} \vdash A$ or $\mathbf{T} \vdash \neg A$. In this sense then, $\mathbf{T}$ would have settled every mathematical question: it would either prove that it's true or that it's false. If Hilbert had been right, it would also have turned out that mathematics is **decidable**. That's because any axiomatizable theory is **computably enumerable**, i.e., there is a computable function that lists all its theorems. We can test if a sentence A is a theorem by listing all of them until we find A (in which it is a theorem) or $\neg A$ (in which case it isn't). Alas, Hilbert was wrong. Gödel proved that no axiomatizable, consistent theory that is "strong enough" is complete. That's the **first incompleteness theorem**. The requirement that the theory be "strong enough" amounts to it representing all computable functions and relations. Specifically, the very weak theory $\mathbf{Q}$ satisfies this property, and any theory that is at least as strong as $\mathbf{Q}$ also does. He also showed—that is the **second incompleteness theorem**—that the sentence that expresses the consistency of the theory is itself undecidable in it, i.e., the theory proves neither it nor its negation. So Hilbert's further aim of finding "finitary" consistency proof of all of mathematics cannot be realized. For any finitary consistency proof would, presumably, be formalizable in a theory that captures all of mathematics. Finally, we established that theories that represent all computable functions and relations are not **decidable**. Note that although axomatizability and completeness implies decidability, incompleteness does not imply undecidability. So this result shows that the second of Hilbert's goals, namely that there be a procedure that decides if $\mathbf{T} \vdash A$ or not, can also not be achieved, at least not for theories at least as strong as $\mathbf{Q}$.

Problems

Problem 1.1. Show that $\mathbf{TA} = \{A : \mathfrak{N} \vDash A\}$ is not axiomatizable. You may assume that $\mathbf{TA}$ represents all decidable properties.

CHAPTER 2

Recursive Functions

2.1 Introduction

In order to develop a mathematical theory of computability, one has to, first of all, develop a *model* of computability. We now think of computability as the kind of thing that computers do, and computers work with symbols. But at the beginning of the development of theories of computability, the paradigmatic example of computation was *numerical* computation. Mathematicians were always interested in number-theoretic functions, i.e., functions $f : \mathbb{N}^n \to \mathbb{N}$ that can be computed. So it is not surprising that at the beginning of the theory of computability, it was such functions that were studied. The most familiar examples of computable numerical functions, such as addition, multiplication, exponentiation (of natural numbers) share an interesting feature: they can be defined *recursively*. It is thus quite natural to attempt a general definition of *computable function* on the basis of recursive definitions. Among the many possible ways to define number-theoretic functions recursively, one particulalry simple pattern of definition here becomes central: so-called *primitive recursion*.

In addition to computable functions, we might be interested

in computable sets and relations. A set is computable if we can compute the answer to whether or not a given number is an element of the set, and a relation is computable iff we can compute whether or not a tuple $\langle n_1, \ldots, n_k \rangle$ is an element of the relation. By considering the *characteristic function* of a set or relation, discussion of computable sets and relations can be subsumed under that of computable functions. Thus we can define primitive recursive relations as well, e.g., the relation "n evenly divides m" is a primitive recursive relation.

Primitive recursive functions—those that can be defined using just primitive recursion—are not, however, the only computable number-theoretic functions. Many generalizations of primitive recursion have been considered, but the most powerful and widely-accepted additional way of computing functions is by unbounded search. This leads to the definition of *partial recursive functions*, and a related definition to *general recursive functions*. General recursive functions are computable and total, and the definition characterizes exactly the partial recursive functions that happen to be total. Recursive functions can simulate every other model of computation (Turing machines, lambda calculus, etc.) and so represent one of the many accepted models of computation.

2.2 Primitive Recursion

A characteristic of the natural numbers is that every natural number can be reached from 0 by applying the successor operation $+1$ finitely many times—any natural number is either 0 or the successor of ... the successor of 0. One way to specify a function $f \colon \mathbb{N} \to \mathbb{N}$ that makes use of this fact is this: (a) specify what the value of f is for argument 0, and (b) also specify how to, given the value of $f(x)$, compute the value of $f(x+1)$. For (a) tells us directly what $f(0)$ is, so f is defined for 0. Now, using the instruction given by (b) for $x = 0$, we can compute $f(1) = f(0+1)$ from $f(0)$. Using the same instructions for $x = 1$, we compute $f(2) = f(1+1)$ from $f(1)$, and so on. For every natural number x, we'll eventually reach the step where we define $f(x)$ from $f(x+1)$, and so $f(x)$ is defined for all $x \in \mathbb{N}$.

For instance, suppose we specify $h\colon \mathbb{N} \to \mathbb{N}$ by the following two equations:

$$\begin{aligned} h(0) &= 1 \\ h(x+1) &= 2 \cdot h(x) \end{aligned}$$

If we already know how to multiply, then these equations give us the information required for (a) and (b) above. Successively the second equation, we get that

$$\begin{aligned} h(1) &= 2 \cdot h(0) = 2, \\ h(2) &= 2 \cdot h(1) = 2 \cdot 2, \\ h(3) &= 2 \cdot h(2) = 2 \cdot 2 \cdot 2, \\ &\vdots \end{aligned}$$

We see that the function h we have specified is $h(x) = 2^x$.

The characteristic feature of the natural numbers guarantees that there is only one function d that meets these two criteria. A pair of equations like these is called a *definition by primitive recursion* of the function d. It is so-called because we define f "recursively," i.e., the definition, specifically the second equation, involves f itself on the right-hand-side. It is "primitive" because in defining $f(x+1)$ we only use the value $f(x)$, i.e., the immediately preceding value. This is the simplest way of defining a function on $\mathbb{N}$ recursively.

We can define even more fundamental functions like addition and multiplication by primitive recursion. In these cases, however, the functions in question are 2-place. We fix one of the argument places, and use the other for the recursion. E.g, to define $\operatorname{add}(x, y)$ we can fix x and define the value first for $y = 0$ and then for $y + 1$ in terms of y. Since x is fixed, it will appear on the left and on the right side of the defining equations.

$$\operatorname{add}(x, 0) = x$$

$$\text{add}(x, y + 1) = \text{add}(x, y) + 1$$

These equations specify the value of add for all x *and* y. To find add(2, 3), for instance, we apply the defining equations for $x = 2$, using the first to find $\text{add}(2, 0) = 2$, then using the second to successively find $\text{add}(2, 1) = 2 + 1 = 3$, $\text{add}(2, 2) = 3 + 1 = 4$, $\text{add}(2, 3) = 4 + 1 = 5$.

In the definition of add we used + on the right-hand-side of the second equation, but only to add 1. In other words, we used the successor function $\text{succ}(z) = z + 1$ and applied it to the previous value $\text{add}(x, y)$ to define $\text{add}(x, y + 1)$. So we can think of the recursive definition as given in terms of a single function which we apply to the previous value. However, it doesn't hurt—and sometimes is necessary—to allow the function to depend not just on the previous value but also on x and y. Consider:

$$\begin{aligned}\text{mult}(x, 0) &= 0\\ \text{mult}(x, y + 1) &= \text{add}(\text{mult}(x, y), x)\end{aligned}$$

This is a primitive recursive definition of a function mult by applying the function add to both the preceding value $\text{mult}(x, y)$ and the first argument x. It also defines the function $\text{mult}(x, y)$ for all arguments x and y. For instance, mult(2, 3) is determined by successively computing mult(2, 0), mult(2, 1), mult(2, 2), and mult(2, 3):

$$\begin{aligned}&\text{mult}(2, 0) = 0\\ &\text{mult}(2, 1) = \text{mult}(2, 0 + 1) = \text{add}(\text{mult}(2, 0), 2) = \text{add}(0, 2) = 2\\ &\text{mult}(2, 2) = \text{mult}(2, 1 + 1) = \text{add}(\text{mult}(2, 1), 2) = \text{add}(2, 2) = 4\\ &\text{mult}(2, 3) = \text{mult}(2, 2 + 1) = \text{add}(\text{mult}(2, 2), 2) = \text{add}(4, 2) = 6\end{aligned}$$

The general pattern then is this: to give a primitive recursive definition of a function $h(x_0, \ldots, x_{k-1}, y)$, we provide two equations. The first defines the value of $h(x_0, \ldots, x_{k-1}, 0)$ without reference to f. The second defines the value of $h(x_0, \ldots, x_{k-1}, y + 1)$ in terms of $h(x_0, \ldots, x_{k-1}, y)$, the other arguments x_0, …, x_{k-1}, and y. Only the immediately preceding value of h may be used in that second equation. If we think of the operations given by the right-hand-sides of these two equations as themselves being functions f and g, then the pattern to define a new function h by primitive recursion is this:

$$h(x_0, \ldots, x_{k-1}, 0) = f(x_0, \ldots, x_{k-1})$$
$$h(x_0, \ldots, x_{k-1}, y + 1) = g(x_0, \ldots, x_{k-1}, y, h(x_0, \ldots, x_{k-1}, y))$$

In the case of add, we have $k = 0$ and $f(x_0) = x_0$ (the identity function), and $g(x_0, y, z) = z + 1$ (the 3-place function that returns the successor of its third argument):

$$\text{add}(x_0, 0) = f(x_0) = x_0$$
$$\text{add}(x_0, y + 1) = g(x_0, y, \text{add}(x_0, y)) = \text{succ}(\text{add}(x_0, y))$$

In the case of mult, we have $f(x_0) = 0$ (the constant function always returning 0) and $g(x_0, y, z) = \text{add}(z, x_0)$ (the 3-place function that returns the sum of its last and first argument):

$$\text{mult}(x_0, 0) = f(x_0) = 0$$
$$\text{mult}(x_0, y + 1) = g(x_0, y, \text{mult}(x_0, y)) = \text{add}(\text{mult}(x_0, y), x_0)$$

2.3 Composition

If f and g are two one-place functions of natural numbers, we can compose them: $h(x) = g(f(x)$. The new function $h(x)$ is then defined by ***composition*** from the functions f and g. We'd like to generalize this to functions of more than one argument.

Here's one way of doing this: suppose f is a k-place function, and g_0, …, g_{k-1} are k functions which are all n-place. Then we can define a new n-place function h as follows:

$$h(x_0, \ldots, x_{n-1}) = f(g_0(x_0, \ldots, x_{n-1}), \ldots, g_{k-1}(x_0, \ldots, x_{n-1}))$$

If f and all g_i are computable, so is h: To compute $h(x_0, \ldots, x_{n-1})$, first compute the values $y_i = g_i(x_0, \ldots, x_{n-1})$ for each $i = 0, \ldots, k-$

1. Then feed these values into f to compute $h(x_0, \ldots, x_{k-1}) = f(y_0, \ldots, y_{k-1})$.

This may seem like an overly restrictive characterization of what happens when we compute a new function using some existing ones. For one thing, sometimes we do not use all the arguments of a function, as when we defined $g(x, y, z) = \mathrm{succ}(z)$ for use in the primitive recursive definition of add. Suppose we are allowed use of the following functions:

$$P_i^n(x_0, \ldots, x_{n-1}) = x_i$$

The functions P_i^k are called *projection* functions: P_i^n is an n-place function. Then g can be defined by

$$g(x, y, z) = \mathrm{succ}(P_2^3).$$

Here the role of f is played by the 1-place function succ, so $k = 1$. And we have one 3-place function P_2^3 which plays the role of g_0. The result is a 3-place function that returns the successor of the third argument.

The projection functions also allow us to define new functions by reordering or identifying arguments. For instance, the function $h(x) = \mathrm{add}(x, x)$ can be defined by

$$h(x_0) = \mathrm{add}(P_0^1(x_0), P_0^1(x_0)).$$

Here $k = 2$, $n = 1$, the role of $f(y_0, y_1)$ is played by add, and the roles of $g_0(x_0)$ and $g_1(x_0)$ are both played by $P_0^1(x_0)$, the one-place projection function (aka the identity function).

If $f(y_0, y_1)$ is a function we already have, we can define the function $h(x_0, x_1) = f(x_1, x_0)$ by

$$h(x_0, x_1) = f(P_1^2(x_0, x_1), P_0^2(x_0, x_1)).$$

Here $k = 2$, $n = 2$, and the roles of g_0 and g_1 are played by P_1^2 and P_0^2, respectively.

You may also worry that $g_0, \ldots, g_{k-1}$ are all required to have the same arity n. (Remember that the *arity* of a function is the number of arguments; an n-place function has arity n.) But adding the projection functions provides the desired flexibility. For example, suppose f and g are 3-place functions and h is the 2-place function defined by

$$h(x, y) = f(x, g(x, x, y), y).$$

The definition of h can be rewritten with the projection functions, as

$$h(x, y) = f(P^2_0(x, y), g(P^2_0(x, y), P^2_0(x, y), P^2_1(x, y)), P^2_1(x, y)).$$

Then h is the composition of f with P^2_0, l, and P^2_1, where

$$l(x, y) = g(P^2_0(x, y), P^2_0(x, y), P^2_1(x, y)),$$

i.e., l is the composition of g with P^2_0, P^2_0, and P^2_1.

2.4 Primitive Recursion Functions

Let us record again how we can define new functions from existing ones using primitive recursion and composition.

Definition 2.1. Suppose f is a k-place function ($k \geq 1$) and g is a $(k + 2)$-place function. The function defined by *primitive recursion from f and g* is the $(k + 1)$-place function h defined by the equations

$$\begin{aligned} h(x_0, \ldots, x_{k-1}, y) &= f(x_0, \ldots, x_{k-1}) \\ h(x_0, \ldots, x_{k-1}, y + 1) &= g(x_0, \ldots, x_{k-1}, y, h(x_0, \ldots, x_{k-1}, y)) \end{aligned}$$

Definition 2.2. Suppose f is a k-place function, and $g_0, \ldots, g_{k-1}$ are k functions which are all n-place. The function defined by *composition from f and $g_0, \ldots, g_{k-1}$* is the n-place function h defined by

$$h(x_0, \ldots, x_{n-1}) = f(g_0(x_0, \ldots, x_{n-1}), \ldots, g_{k-1}(x_0, \ldots, x_{n-1})).$$

In addition to succ and the projection functions

$$P_i^n(x_0, \ldots, x_{n-1}) = x_i,$$

for each natural number n and $i < n$, we will include among the primitive recursive functions the function $\text{zero}(x) = 0$.

Definition 2.3. The set of primitive recursive functions is the set of functions from $\mathbb{N}^n$ to $\mathbb{N}$, defined inductively by the following clauses:

1. zero is primitive recursive.
2. succ is primitive recursive.
3. Each projection function P_i^n is primitive recursive.
4. If f is a k-place primitive recursive function and g_0, $\ldots$, g_{k-1} are n-place primitive recursive functions, then the composition of f with g_0, $\ldots$, g_{k-1} is primitive recursive.
5. If f is a k-place primitive recursive function and g is a $k+2$-place primitive recursive function, then the function defined by primitive recursion from f and g is primitive recursive.

Put more concisely, the set of primitive recursive functions is the smallest set containing zero, succ, and the projection functions P_j^n, and which is closed under composition and primitive recursion.

Another way of describing the set of primitive recursive functions is by defining it in terms of "stages." Let S_0 denote the set of starting functions: zero, succ, and the projections. These are the primitive recursive functions of stage 0. Once a stage S_i has been defined, let S_{i+1} be the set of all functions you get by applying a single instance of composition or primitive recursion to functions already in S_i. Then

$$S = \bigcup_{i \in \mathbb{N}} S_i$$

is the set of all primitive recursive functions

Let us verify that add is a primitive recursive function.

Proposition 2.4. *The addition function* $\mathrm{add}(x, y) = x + y$ *is primitive recursive.*

Proof. We already have a primitive recursive definition of add in terms of two functions f and g which matches the format of Definition 2.1:

$$\begin{aligned}
\mathrm{add}(x_0, 0) &= f(x_0) = x_0 \\
\mathrm{add}(x_0, y+1) &= g(x_0, y, \mathrm{add}(x_0, y)) = \mathrm{succ}(\mathrm{add}(x_0, y))
\end{aligned}$$

So add is primitive recursive provided f and g are as well. $f(x_0) = x_0 = P^1_0(x_0)$, and the projection functions count as primitive recursive, so f is primitive recursive. The function g is the three-place function $g(x_0, y, z)$ defined by

$$g(x_0, y, z) = \mathrm{succ}(z).$$

This does not yet tell us that g is primitive recursive, since g and succ are not quite the same function: succ is one-place, and g has to be three-place. But we can define g "officially" by composition as

$$g(x_0, y, z) = \mathrm{succ}(P^3_2(x_0, y, z))$$

Since succ and P^3_2 count as primitive recursive functions, g does as well, since it can be defined by composition from primitive recursive functions. □

Proposition 2.5. *The multiplication function* $\text{mult}(x, y) = x \cdot y$ *is primitive recursive.*

Proof. Exercise. □

Example 2.6. Here's our very first example of a primitive recursive definition:

$$h(0) = 1$$
$$h(y + 1) = 2 \cdot h(y).$$

This function cannot fit into the form required by Definition 2.1, since $k = 0$. The definition also involves the constants 1 and 2. To get around the first problem, let's introduce a dummy argument and define the function h':

$$h'(x_0, 0) = f(x_0) = 1$$
$$h'(x_0, y + 1) = g(x_0, y, h'(x_0, y)) = 2 \cdot h'(x_0, y).$$

The function $f(x_0) = 1$ can be defined from succ and zero by composition: $f(x_0) = \text{succ}(\text{zero}(x_0))$. The function g can be defined by composition from $g'(z) = 2 \cdot z$ and projections:

$$g(x_0, y, z) = g'(P^3_2(x_0, y, z))$$

and g' in turn can be defined by composition as

$$g'(z) = \text{mult}(g''(z), P^1_0(z))$$

and

$$g''(z) = \text{succ}(f(z)),$$

where f is as above: $f(z) = \text{succ}(\text{zero}(z))$. Now that we have h' we can use composition again to let $h(y) = h'(P^1_0(y), P^1_0(y))$. This shows that h can be defined from the basic functions using a sequence of compositions and primitive recursions, so h is primitive recursive.

2.5 Primitive Recursion Notations

One advantage to having the precise inductive description of the primitive recursive functions is that we can be systematic in describing them. For example, we can assign a "notation" to each such function, as follows. Use symbols zero, succ, and P^n_i for zero, successor, and the projections. Now suppose f is defined by composition from a k-place function h and n-place functions g_0, ..., g_{k-1}, and we have assigned notations H, G_0, ..., G_{k-1} to the latter functions. Then, using a new symbol $\text{Comp}_{k,n}$, we can denote the function f by $\text{Comp}_{k,n}[H, G_0, \ldots, G_{k-1}]$. For the functions defined by primitive recursion, we can use analogous notations of the form $\text{Rec}_k[G, H]$, where $k + 1$ is the arity of the function being defined. With this setup, we can denote the addition function by

$$\text{Rec}_2[P^1_0, \text{Comp}_{1,3}[\text{succ}, P^3_2]].$$

Having these notations sometimes proves useful.

2.6 Primitive Recursive Functions are Computable

Suppose a function h is defined by primitive recursion

$$\begin{aligned} h(\vec{x}, 0) &= f(\vec{x}) \\ h(\vec{x}, y) &= g(\vec{x}, y, h(\vec{x}, y)) \end{aligned}$$

and suppose the functions f and g are computable. (We use $\vec{x}$ to abbreviate $x_0, \ldots, x_{k-1}$.) Then $h(\vec{x}, 0)$ can obviously be computed, since it is just $f(\vec{x})$ which we assume is computable. $h(\vec{x}, 1)$ can then also be computed, since $1 = 0 + 1$ and so $h(\vec{x}, 1)$ is just

$$h(\vec{x}, 1) = g(\vec{x}, 0, h(\vec{x}, 0)) = g(\vec{x}, 0, f(\vec{x})).$$

We can go on in this way and compute

$$h(\vec{x}, 2) = g(\vec{x}, 1, h(\vec{x}, 1)) = g(\vec{x}, 1, g(\vec{x}, 0, f(\vec{x})))$$

$$h(\vec{x}, 3) = g(\vec{x}, 2, h(\vec{x}, 2)) = g(\vec{x}, 2, g(\vec{x}, 1, g(\vec{x}, 0, f(\vec{x}))))$$
$$h(\vec{x}, 4) = g(\vec{x}, 3, h(\vec{x}, 3)) = g(\vec{x}, 3, g(\vec{x}, 2, g(\vec{x}, 1, g(\vec{x}, 0, f(\vec{x})))))$$
$$\vdots$$

Thus, to compute $h(\vec{x}, y)$ in general, successively compute $h(\vec{x}, 0)$, $h(\vec{x}, 1)$, …, until we reach $h(\vec{x}, y)$.

Thus, a primitive recursive definition yields a new computable function if the functions f and g are computable. Composition of functions also results in a computable function if the functions f and g_i are computable.

Since the basic functions zero, succ, and P^n_i are computable, and composition and primitive recursion yield computable functions from computable functions, this means that every primitive recursive function is computable.

2.7 Examples of Primitive Recursive Functions

We already have some examples of primitive recursive functions: the addition and multiplication functions add and mult. The identity function $\mathrm{id}(x) = x$ is primitive recursive, since it is just P^1_0. The constant functions $\mathrm{const}_n(x) = n$ are primitive recursive since they can be defined from zero and succ by successive composition. This is useful when we want to use constants in primitive recursive definitions, e.g., if we want to define the function $f(x) = 2 \cdot x$ can obtain it by composition from $\mathrm{const}_n(x)$ and multiplication as $f(x) = \mathrm{mult}(\mathrm{const}_2(x), P^1_0(x))$. We'll make use of this trick from now on.

Proposition 2.7. *The exponentiation function* $\exp(x, y) = x^y$ *is primitive recursive.*

Proof. We can define exp primitive recursively as

$$\exp(x, 0) = 1$$

$$\exp(x, y+1) = \text{mult}(x, \exp(x, y)).$$

Strictly speaking, this is not a recursive definition from primitive recursive functions. Officially, though, we have:

$$\begin{aligned} \exp(x, 0) &= f(x) \\ \exp(x, y+1) &= g(x, y, \exp(x, y)). \end{aligned}$$

where

$$\begin{aligned} f(x) &= \text{succ}(\text{zero}(x)) = 1 \\ g(x, y, z) &= \text{mult}(P^3_0(x, y, z), P^3_2(x, y, z) = x \cdot z \end{aligned}$$

and so f and g are defined from primitive recursive functions by composition. □

Proposition 2.8. *The predecessor function* pred(y) *defined by*

$$\text{pred}(y) = \begin{cases} 0 & \textit{if } y = 0 \\ y - 1 & \textit{otherwise} \end{cases}$$

is primitive recursive.

Proof. Note that

$$\begin{aligned} \text{pred}(0) &= 0 \text{ and} \\ \text{pred}(y+1) &= y. \end{aligned}$$

This is almost a primitive recursive definition. It does not, strictly speaking, fit into the pattern of definition by primitive recursion, since that pattern requires at least one extra argument x. It is also odd in that it does not actually use pred(y) in the definition of pred($y + 1$). But we can first define $\text{pred}'(x, y)$ by

$$\begin{aligned} \text{pred}'(x, 0) &= \text{zero}(x) = 0, \\ \text{pred}'(x, y+1) &= P^3_1(x, y, \text{pred}'(x, y)) = y. \end{aligned}$$

and then define pred from it by composition, e.g., as $\text{pred}(x) = \text{pred}'(\text{zero}(x), P^1_0(x))$. □

Proposition 2.9. *The factorial function* $\text{fac}(x) = x! = 1 \cdot 2 \cdot 3 \cdot \cdots \cdot x$ *is primitive recursive.*

Proof. The obvious primitive recursive definition is

$$\text{fac}(0) = 1$$
$$\text{fac}(y+1) = \text{fac}(y) \cdot (y+1).$$

Officially, we have to first define a two-place function h

$$h(x, 0) = \text{const}_1(x)$$
$$h(x, y) = g(x, y, h(x, y))$$

where $g(x, y, z) = \text{mult}(P^3_2(x, y, z), \text{succ}(P^3_1(x, y, z))$ and then let

$$\text{fac}(y) = h(P^1_0(y), P^1_0(y))$$

From now on we'll be a bit more laissez-faire and not give the official definitions by composition and primitive recursion. □

Proposition 2.10. *Truncated subtraction,* $x \dot{-} y$, *defined by*

$$x \dot{-} y = \begin{cases} 0 & \text{if } x > y \\ x - y & \text{otherwise} \end{cases}$$

is primitive recursive.

Proof. We have:

$$x \dot{-} 0 = x$$
$$x \dot{-} (y+1) = \text{pred}(x \dot{-} y)$$

□

Proposition 2.11. *The distance between x and y, $|x-y|$, is primitive recursive.*

Proof. We have $|x-y| = (x \dot{-} y) + (y \dot{-} x)$, so the distance can be defined by composition from $+$ and $\dot{-}$, which are primitive recursive. □

Proposition 2.12. *The maximum of x and y, $\max(x, y)$, is primitive recursive.*

Proof. We can define $\max(x, y)$ by composition from $+$ and $\dot{-}$ by

$$\max(x, y) = x + (y \dot{-} x).$$

If x is the maximum, i.e., $x \geq y$, then $y \dot{-} x = 0$, so $x + (y \dot{-} x) = x + 0 = x$. If y is the maximum, then $y \dot{-} x = y - x$, and so $x + (y \dot{-} x) = x + (y - x) = y$. □

Proposition 2.13. *The minimum of x and y, $\min(x, y)$, is primitive recursive.*

Proof. Exercise. □

Proposition 2.14. *The set of primitive recursive functions is closed under the following two operations:*

1. *Finite sums: if $f(\vec{x}, z)$ is primitive recursive, then so is the function*

$$g(\vec{x}, y) = \sum_{z=0}^{y} f(\vec{x}, z).$$

2. *Finite products: if $f(\vec{x}, z)$ is primitive recursive, then so is the*

function

$$h(\vec{x}, y) = \prod_{z=0}^{y} f(\vec{x}, z).$$

Proof. For example, finite sums are defined recursively by the equations

$$g(\vec{x}, 0) = f(\vec{x}, 0)$$
$$g(\vec{x}, y+1) = g(\vec{x}, y) + f(\vec{x}, y+1).$$ □

2.8 Primitive Recursive Relations

Definition 2.15. A relation $R(\vec{x})$ is said to be primitive recursive if its characteristic function,

$$\chi_R(\vec{x}) = \begin{cases} 1 & \text{if } R(\vec{x}) \\ 0 & \text{otherwise} \end{cases}$$

is primitive recursive.

In other words, when one speaks of a primitive recursive relation $R(\vec{x})$, one is referring to a relation of the form $\chi_R(\vec{x}) = 1$, where χ_R is a primitive recursive function which, on any input, returns either 1 or 0. For example, the relation IsZero(x), which holds if and only if $x = 0$, corresponds to the function χ_{IsZero}, defined using primitive recursion by

$$\chi_{\text{IsZero}}(0) = 1, \quad \chi_{\text{IsZero}}(x+1) = 0.$$

It should be clear that one can compose relations with other primitive recursive functions. So the following are also primitive recursive:

1. The equality relation, $x = y$, defined by IsZero($|x - y|$)
2. The less-than relation, $x \leq y$, defined by IsZero($x \dot{-} y$)

Proposition 2.16. *The set of primitive recursive relations is closed under boolean operations, that is, if $P(\vec{x})$ and $Q(\vec{x})$ are primitive, so are*

1. $\neg R(\vec{x})$
2. $P(\vec{x}) \land Q(\vec{x})$
3. $P(\vec{x}) \lor Q(\vec{x})$
4. $P(\vec{x}) \rightarrow Q(\vec{x})$

Proof. Suppose $P(\vec{x})$ and $Q(\vec{x})$ are primitive recursive, i.e., their characteristic functions χ_P and χ_Q are. We have to show that the characteristic functions of $\neg R(\vec{x})$, etc., are also primitive recursive.

$$\chi_{\neg P}(\vec{x}) = \begin{cases} 0 & \text{if } \chi_P(\vec{x}) = 1 \\ 1 & \text{otherwise} \end{cases}$$

We can define $\chi_{\neg P}(\vec{x})$ as $1 \dot{-} \chi_P(\vec{x})$.

$$\chi_{P \land Q}(\vec{x}) = \begin{cases} 1 & \text{if } \chi_P(\vec{x}) = \chi_Q(\vec{x}) = 1 \\ 0 & \text{otherwise} \end{cases}$$

We can define $\chi_{P \land Q}(\vec{x})$ as $\chi_P(\vec{x}) \cdot \chi_Q(\vec{x})$ or as $\min(\chi_P(\vec{x}), \chi_Q(\vec{x}))$.

Similarly, $\chi_{P \lor Q}(\vec{x}) = \max(\chi_P(\vec{x}), \chi_Q(\vec{x}))$ and $\chi_{P \lor Q}(\vec{x}) = \max(1 \dot{-} \chi_P(\vec{x}), \chi_Q(\vec{x}))$. □

Proposition 2.17. *The set of primitive recursive relations is closed under bounded quantification, i.e., if $R(\vec{x}, z)$ is a primitive recursive relation, then so are the relations $(\forall z < y)\ R(\vec{x}, z)$ and $(\exists z < y)\ R(\vec{x}, z)$.*

($(\forall z < y)\ R(\vec{x}, z)$ holds of $\vec{x}$ and y if and only if $R(\vec{x}, z)$ holds for every z less than y, and similarly for $(\exists z < y)\ R(\vec{x}, z)$.)

Proof. By convention, we take $(\forall z < 0)\ R(\vec{x}, z)$ to be true (for the trivial reason that there are no z less than 0) and $(\exists z < 0)\ R(\vec{x}, z)$

to be false. A universal quantifier functions just like a finite product or iterated minimum, i.e., if $P(\vec{x}, y) \Leftrightarrow (\forall z < y)\ R(\vec{x}, z)$ then $\chi_P(\vec{x}, y)$ can be defined by

$$\chi_P(\vec{x}, 0) = 1$$
$$\chi_P(\vec{x}, y + 1) = \min(\chi_P(\vec{x}, y), \chi_R(\vec{x}, y + 1))).$$

Bounded existential quantification can similarly be defined using max. Alternatively, it can be defined from bounded universal quantification, using the equivalence $(\exists z < y)\ R(\vec{x}, z) \leftrightarrow \neg(\forall z < y)\ \neg R(\vec{x}, z)$. Note that, for example, a bounded quantifier of the form $(\exists x \leq y)\ \ldots x \ldots$ is equivalent to $(\exists x < y + 1)\ \ldots x \ldots$. □

Another useful primitive recursive function is the conditional function, $\mathrm{cond}(x, y, z)$, defined by

$$\mathrm{cond}(x, y, z) = \begin{cases} y & \text{if } x = 0 \\ z & \text{otherwise.} \end{cases}$$

This is defined recursively by

$$\mathrm{cond}(0, y, z) = y, \quad \mathrm{cond}(x + 1, y, z) = z.$$

One can use this to justify definitions of primitive recursive functions by cases from primitive recursive relations:

Proposition 2.18. *If $g_0(\vec{x}), \ldots, g_m(\vec{x})$ are functions, and $R_1(\vec{x}), \ldots, R_{m-1}(\vec{x})$ are primitive recursive relations, then the function f defined by*

$$f(\vec{x}) = \begin{cases} g_0(\vec{x}) & \textit{if } R_0(\vec{x}) \\ g_1(\vec{x}) & \textit{if } R_1(\vec{x}) \textit{ and not } R_0(\vec{x}) \\ \vdots & \\ g_{m-1}(\vec{x}) & \textit{if } R_{m-1}(\vec{x}) \textit{ and none of the previous hold} \\ g_m(\vec{x}) & \textit{otherwise} \end{cases}$$

is also primitive recursive.

Proof. When $m = 1$, this is just the function defined by

$$f(\vec{x}) = \text{cond}(\chi_{\neg R_0}(\vec{x}), g_0(\vec{x}), g_1(\vec{x})).$$

For m greater than 1, one can just compose definitions of this form. □

2.9 Bounded Minimization

It is often useful to define a function as the least number satisfying some property or relation P. If P is decidable, we can compute this function simply by trying out all the possible numbers, 0, 1, 2, ..., until we find the least one satisfying P. This kind of unbounded search takes us out of the realm of primitive recursive functions. However, if we're only interested in the least number *less than some independently given bound*, we stay primitive recursive. In other words, and a bit more generally, suppose we have a primitive recursive relation $R(x, z)$. Consider the function that maps x and y to the least $z < y$ such that $R(x, z)$. It, too, can be computed, by testing whether $R(x, 0)$, $R(x, 1)$, ..., $R(x, y - 1)$. But why is it primitive recursive?

Proposition 2.19. *If $R(\vec{x}, z)$ is primitive recursive, so is the function $m_R(\vec{x}, y)$ which returns the least z less than y such that $R(\vec{x}, z)$ holds, if there is one, and y otherwise. We will write the function m_R as*

$$(\min z < y)\, R(\vec{x}, z),$$

Proof. Note than there can be no $z < 0$ such that $R(\vec{x}, z)$ since there is no $z < 0$ at all. So $m_R(\vec{x}, 0) = 0$.

In case the bound is of the form $y + 1$ we have three cases: (a) There is a $z < y$ such that $R(\vec{x}, z)$, in which case $m_R(\vec{x}, y + 1) = m_R(\vec{x}, y)$. (b) There is no such $z < y$ but $R(\vec{x}, y)$ holds, then

$m_R(\vec{x}, y+1) = y$. (c) There is no $z < y+1$ such that $R(\vec{x}, z)$, then $m_R(\vec{z}, y+1) = y+1$. So,

$$m_R(\vec{x}, 0) = 0$$

$$m_R(\vec{x}, y+1) = \begin{cases} m_R(\vec{x}, y) & \text{if } m_R(\vec{x}, y) \neq y \\ y & \text{if } m_R(\vec{x}, y) = y \text{ and } R(\vec{x}, y) \\ y+1 & \text{otherwise.} \end{cases}$$

Note that there is a $z < y$ such that $R(\vec{x}, z)$ iff $m_R(\vec{x}, y) \neq y$. □

2.10 Primes

Bounded quantification and bounded minimization provide us with a good deal of machinery to show that natural functions and relations are primitive recursive. For example, consider the relation "x divides y", written $x \mid y$. The relation $x \mid y$ holds if division of y by x is possible without remainder, i.e., if y is an integer multiple of x. (If it doesn't hold, i.e., the remainder when dividing x by y is > 0, we write $x \nmid y$.) In other words, $x \mid y$ iff for some z, $x \cdot z = y$. Obviously, any such z, if it exists, must be $\leq y$. So, we have that $x \mid y$ iff for some $z \leq y$, $x \cdot z = y$. We can define the relation $x \mid y$ by bounded existential quantification from $=$ and multiplication by

$$x \mid y \Leftrightarrow (\exists z \leq y)\ (x \cdot z) = y.$$

We've thus shown that $x \mid y$ is primitive recursive.

A natural number x is *prime* if it is neither 0 nor 1 and is only divisible by 1 and itself. In other words, prime numbers are such that, whenever $y \mid x$, either $y = 1$ or $y = x$. To test if x is prime, we only have to check if $y \mid x$ for all $y \leq x$, since if $y > x$, then automatically $y \nmid x$. So, the relation Prime(x), which holds iff x is prime, can be defined by

$$\text{Prime}(x) \Leftrightarrow x \geq 2 \land (\forall y \leq x)\ (y \mid x \rightarrow y = 1 \lor y = x)$$

and is thus primitive recursive.

The primes are 2, 3, 5, 7, 11, etc. Consider the function $p(x)$ which returns the xth prime in that sequence, i.e., $p(0) = 2$, $p(1) = 3$, $p(2) = 5$, etc. (For convenience we will often write $p(x)$ as p_x ($p_0 = 2$, $p_1 = 3$, etc.)

If we had a function nextPrime(x), which returns the first prime number larger than x, p can be easily defined using primitive recursion:

$$p(0) = 2$$
$$p(x+1) = \text{nextPrime}(p(x))$$

Since nextPrime(x) is the least y such that $y > x$ and y is prime, it can be easily computed by unbounded search. But it can also be defined by bounded minimization, thanks to a result due to Euclid: there is always a prime number between x and $x\,! + 1$.

$$\text{nextPrime(x)} = (\min y \leq x\,! + 1)\,(y > x \land \text{Prime}(y)).$$

This shows, that nextPrime(x) and hence $p(x)$ are (not just computable but) primitive recursive.

(If you're curious, here's a quick proof of Euclid's theorem. Suppose p_n is the largest prime $\leq x$ and consider the product $p = p_0 \cdot p_1 \cdot \dots \cdot p_n$ of all primes $\leq x$. Either $p+1$ is prime or there is a prime between x and $p+1$. Why? Suppose $p+1$ is not prime. Then some prime number $q \mid p+1$ where $q < p+1$. None of the primes $\leq x$ divide $p+1$. (By definition of p, each of the primes $p_i \leq x$ divides p, i.e., with remainder 0. So, each of the primes $p_i \leq x$ divides $p+1$ with remainder 1, and so $p_i \nmid p+1$.) Hence, q is a prime $> x$ and $< p+1$. And $p \leq x\,!$, so there is a prime $> x$ and $\leq x\,! + 1$.)

2.11 Sequences

The set of primitive recursive functions is remarkably robust. But we will be able to do even more once we have developed a adequate means of handling *sequences*. We will identify finite sequences of natural numbers with natural numbers in the following way: the sequence $\langle a_0, a_1, a_2, \ldots, a_k \rangle$ corresponds to the number

$$p_0^{a_0+1} \cdot p_1^{a_1+1} \cdot p_2^{a_2+1} \cdot \cdots \cdot p_k^{a_k+1}.$$

We add one to the exponents to guarantee that, for example, the sequences $\langle 2, 7, 3 \rangle$ and $\langle 2, 7, 3, 0, 0 \rangle$ have distinct numeric codes. We can take both 0 and 1 to code the empty sequence; for concreteness, let Λ denote 0.

The reason that this coding of sequences works is the so-called Fundamental Theorem of Arithmetic: every natural number $n \geq 2$ can be written in one and only one way in the form

$$n = p_0^{a_0} \cdot p_1^{a_1} \cdot \cdots \cdot p_k^{a_k}$$

with $a_k \geq 1$. This guarantees that the mapping $\langle\rangle(a_0, \ldots, a_k) = \langle a_0, \ldots, a_k \rangle$ is injective: different sequences are mapped to different numbers; to each number only at most one sequence corresponds.

We'll now show that the operations of determining the length of a sequence, determining its ith element, appending an element to a sequence, and concatenating two sequences, are all primitive recursive.

Proposition 2.20. *The function* $\mathrm{len}(s)$, *which returns the length of the sequence* s, *is primitive recursive.*

Proof. Let $R(i, s)$ be the relation defined by

$$R(i, s) \text{ iff } p_i \mid s \land p_{i+1} \nmid s.$$

R is clearly primitive recursive. Whenever s is the code of a non-empty sequence, i.e.,

$$s = p_0^{a_0+1} \cdot \cdots \cdot p_k^{a_k+1},$$

$R(i, s)$ holds if p_i is the largest prime such that $p_i \mid s$, i.e., $i = k$. The length of s thus is $i+1$ iff p_i is the largest prime that divides s, so we can let

$$\text{len}(s) = \begin{cases} 0 & \text{if } s = 0 \text{ or } s = 1 \\ 1 + (\min i < s)\, R(i, s) & \text{otherwise} \end{cases}$$

We can use bounded minimization, since there is only one i that satisfies $R(s, i)$ when s is a code of a sequence, and if i exists it is less than s itself. □

Proposition 2.21. *The function* append(s, a), *which returns the result of appending a to the sequence s, is primitive recursive.*

Proof. append can be defined by:

$$\text{append}(s, a) = \begin{cases} 2^{a+1} & \text{if } s = 0 \text{ or } s = 1 \\ s \cdot p_{\text{len}(s)}^{a+1} & \text{otherwise.} \end{cases}$$

□

Proposition 2.22. *The function* element(s, i), *which returns the ith element of s (where the initial element is called the 0th), or 0 if i is greater than or equal to the length of s, is primitive recursive.*

Proof. Note that a is the ith element of s iff p_i^{a+1} is the largest power of p_i that divides s, i.e., $p_i^{a+1} \mid s$ but $p_i^{a+2} \nmid s$. So:

$$\text{element}(s, i) = \begin{cases} 0 & \text{if } i \geq \text{len}(s) \\ (\min a < s)\, (p_i^{a+2} \nmid s) & \text{otherwise.} \end{cases}$$

□

Instead of using the official names for the functions defined above, we introduce a more compact notation. We will use $(s)_i$ instead of element(s, i), and $\langle s_0, \dots, s_k \rangle$ to abbreviate

$$\text{append}(\text{append}(\dots \text{append}(\Lambda, s_0) \dots), s_k).$$

Note that if s has length k, the elements of s are $(s)_0, \dots, (s)_{k-1}$.

Proposition 2.23. *The function* concat(s, t), *which concatenates two sequences, is primitive recursive.*

Proof. We want a function concat with the property that

$$\text{concat}(\langle a_0, \ldots, a_k \rangle, \langle b_0, \ldots, b_l \rangle) = \langle a_0, \ldots, a_k, b_0, \ldots, b_l \rangle.$$

We'll use a "helper" function hconcat(s, t, n) which concatenates the first n symbols of t to s. This function can be defined by primitive recursion as follows:

$$\begin{aligned} \text{hconcat}(s, t, 0) &= s \\ \text{hconcat}(s, t, n+1) &= \text{append}(\text{hconcat}(s, t, n), (t)_n) \end{aligned}$$

Then we can define concat by

$$\text{concat}(s, t) = \text{hconcat}(s, t, \text{len}(t)).$$ □

We will write $s \frown t$ instead of concat(s, t).

It will be useful for us to be able to bound the numeric code of a sequence in terms of its length and its largest element. Suppose s is a sequence of length k, each element of which is less than or equal to some number x. Then s has at most k prime factors, each at most p_{k-1}, and each raised to at most $x + 1$ in the prime factorization of s. In other words, if we define

$$\text{sequenceBound}(x, k) = p_{k-1}^{k \cdot (x+1)},$$

then the numeric code of the sequence s described above is at most sequenceBound(x, k).

Having such a bound on sequences gives us a way of defining new functions using bounded search. For example, we can define concat using bounded search. All we need to do is write down a primitive recursive *specification* of the object (number of the concatenated sequence) we are looking for, and a bound on how far to look. The following works:

$$\text{concat}(s, t) = (\min v < \text{sequenceBound}(s + t, \text{len}(s) + \text{len}(t)))$$

$$\begin{aligned}(&\text{len}(v) = \text{len}(s) + \text{len}(t) \land \\ &(\forall i < \text{len}(s))\, ((v)_i = (s)_i) \land \\ &(\forall j < \text{len}(t))\, ((v)_{\text{len}(s)+j} = (t)_j))\end{aligned}$$

Proposition 2.24. *The function* subseq(s, i, n) *which returns the subsequence of s of length n beginning at the ith element, is primitive recursive.*

Proof. Exercise. □

2.12 Trees

Sometimes it is useful to represent trees as natural numbers, just like we can represent sequences by numbers and properties of and operations on them by primitive recursive relations and functions on their codes. We'll use sequences and their codes to do this. A tree can be either a single node (possibly with a label) or else a node (possibly with a label) connected to a number of subtrees. The node is called the *root* of the tree, and the subtrees it is connected to its *immediate subtrees.*

We code trees recursively as a sequence $\langle k, d_1, \ldots, d_k \rangle$, where k is the number of immediate subtrees and d_1, ..., d_k the codes of the immediate subtrees. If the nodes have labels, they can be included after the immediate subtrees. So a tree consisting just of a single node with label l would be coded by $\langle 0, l \rangle$, and a tree consisting of a root (labelled l_1) connected to two single nodes (labelled l_2, l_3) would be coded by $\langle 2, \langle 0, l_2 \rangle, \langle 0, l_3 \rangle, l_1 \rangle$.

Proposition 2.25. *The function* SubtreeSeq(t), *which returns the code of a sequence the elements of which are the codes of all subtrees of the tree with code t, is primitive recursive.*

Proof. First note that ISubtrees(t) = subseq$(t, 1, (t)_0)$ is primitive recursive and returns the codes of the immediate subtrees of a tree t. Now we can define a helper function hSubtreeSeq(t, n)

which computes the sequence of all subtrees which are n nodes remove from the root. The sequence of subtrees of t which is 0 nodes removed from the root—in other words, begins at the root of t—is the sequence consisting just of t. To obtain a sequence of all level $n+1$ subtrees of t, we concatenate the level n subtrees with a sequence consisting of all immediate subtrees of the level n subtrees. To get a list of all these, note that if $f(x)$ is a primitive recursive function returning codes of sequences, then $g_f(s,k) = f((s)_0) \frown \ldots \frown f((s)_k)$ is also primivive recursive:

$$\begin{aligned} g(s,0) &= f((s)_0) \\ g(s,k+1) &= g(s,k) \frown f((s)_{k+1}) \end{aligned}$$

For instance, if s is a sequence of trees, then $h(s) = g_{\text{ISubtrees}}(s, \text{len}(s))$ gives the sequence of the immediate subtrees of the elements of s. We can use it to define hSubtreeSeq by

$$\begin{aligned} \text{hSubtreeSeq}(t,0) &= \langle t \rangle \\ \text{hSubtreeSeq}(t,n+1) &= \text{hSubtreeSeq}(t,n) \frown h(\text{hSubtree}(t,n)). \end{aligned}$$

The maximum level of subtrees in a tree coded by t, i.e., the maximum distance between the root and a leaf node, is bounded by the code t. So a sequence of codes of all subtrees of the tree coded by t is given by hSubtreeSeq(t,t). □

2.13 Other Recursions

Using pairing and sequencing, we can justify more exotic (and useful) forms of primitive recursion. For example, it is often useful to define two functions simultaneously, such as in the following definition:

$$\begin{aligned} h_0(\vec{x},0) &= f_0(\vec{x}) \\ h_1(\vec{x},0) &= f_1(\vec{x}) \\ h_0(\vec{x},y+1) &= g_0(\vec{x},y,h_0(\vec{x},y),h_1(\vec{x},y)) \\ h_1(\vec{x},y+1) &= g_1(\vec{x},y,h_0(\vec{x},y),h_1(\vec{x},y)) \end{aligned}$$

This is an instance of *simultaneous recursion.* Another useful way of defining functions is to give the value of $h(\vec{x}, y+1)$ in terms of *all* the values $h(\vec{x}, 0), \ldots, h(\vec{x}, y)$, as in the following definition:

$$\begin{aligned} h(\vec{x}, 0) &= f(\vec{x}) \\ h(\vec{x}, y+1) &= g(\vec{x}, y, \langle h(\vec{x}, 0), \ldots, h(\vec{x}, y) \rangle). \end{aligned}$$

The following schema captures this idea more succinctly:

$$h(\vec{x}, y) = g(\vec{x}, y, \langle h(\vec{x}, 0), \ldots, h(\vec{x}, y-1) \rangle)$$

with the understanding that the last argument to g is just the empty sequence when y is 0. In either formulation, the idea is that in computing the "successor step," the function h can make use of the entire sequence of values computed so far. This is known as a *course-of-values* recursion. For a particular example, it can be used to justify the following type of definition:

$$h(\vec{x}, y) = \begin{cases} g(\vec{x}, y, h(\vec{x}, k(\vec{x}, y))) & \text{if } k(\vec{x}, y) < y \\ f(\vec{x}) & \text{otherwise} \end{cases}$$

In other words, the value of h at y can be computed in terms of the value of h at *any* previous value, given by k.

You should think about how to obtain these functions using ordinary primitive recursion. One final version of primitive recursion is more flexible in that one is allowed to change the *parameters* (side values) along the way:

$$\begin{aligned} h(\vec{x}, y) &= f(\vec{x}) \\ h(\vec{x}, y+1) &= g(\vec{x}, y, h(k(\vec{x}), y)) \end{aligned}$$

This, too, can be simulated with ordinary primitive recursion. (Doing so is tricky. For a hint, try unwinding the computation by hand.)

2.14 Non-Primitive Recursive Functions

The primitive recursive functions do not exhaust the intuitively computable functions. It should be intuitively clear that we can make a list of all the unary primitive recursive functions, f_0, f_1, f_2, ...such that we can effectively compute the value of f_x on input y; in other words, the function $g(x, y)$, defined by

$$g(x, y) = f_x(y)$$

is computable. But then so is the function

$$\begin{aligned} h(x) &= g(x, x) + 1 \\ &= f_x(x) + 1. \end{aligned}$$

For each primitive recursive function f_i, the value of h and f_i differ at i. So h is computable, but not primitive recursive; and one can say the same about g. This is an "effective" version of Cantor's diagonalization argument.

One can provide more explicit examples of computable functions that are not primitive recursive. For example, let the notation $g^n(x)$ denote $g(g(\ldots g(x)))$, with n g's in all; and define a sequence $g_0, g_1, \ldots$ of functions by

$$\begin{aligned} g_0(x) &= x + 1 \\ g_{n+1}(x) &= g_n^x(x) \end{aligned}$$

You can confirm that each function g_n is primitive recursive. Each successive function grows much faster than the one before; $g_1(x)$ is equal to $2x$, $g_2(x)$ is equal to $2^x \cdot x$, and $g_3(x)$ grows roughly like an exponential stack of x 2's. Ackermann's function is essentially the function $G(x) = g_x(x)$, and one can show that this grows faster than any primitive recursive function.

Let us return to the issue of enumerating the primitive recursive functions. Remember that we have assigned symbolic notations to each primitive recursive function; so it suffices to enumerate notations. We can assign a natural number $\#(F)$ to each notation F, recursively, as follows:

$$\begin{aligned}
\#(0) &= \langle 0 \rangle \\
\#(S) &= \langle 1 \rangle \\
\#(P^n_i) &= \langle 2, n, i \rangle \\
\#(\mathrm{Comp}_{k,l}[H, G_0, \ldots, G_{k-1}]) &= \langle 3, k, l, \#(H), \#(G_0), \ldots, \#(G_{k-1}) \rangle \\
\#(\mathrm{Rec}_l[G, H]) &= \langle 4, l, \#(G), \#(H) \rangle
\end{aligned}$$

Here we are using the fact that every sequence of numbers can be viewed as a natural number, using the codes from the last section. The upshot is that every code is assigned a natural number. Of course, some sequences (and hence some numbers) do not correspond to notations; but we can let f_i be the unary primitive recursive function with notation coded as i, if i codes such a notation; and the constant 0 function otherwise. The net result is that we have an explicit way of enumerating the unary primitive recursive functions.

(In fact, some functions, like the constant zero function, will appear more than once on the list. This is not just an artifact of our coding, but also a result of the fact that the constant zero function has more than one notation. We will later see that one can not computably avoid these repetitions; for example, there is no computable function that decides whether or not a given notation represents the constant zero function.)

We can now take the function $g(x, y)$ to be given by $f_x(y)$, where f_x refers to the enumeration we have just described. How do we know that $g(x, y)$ is computable? Intuitively, this is clear: to compute $g(x, y)$, first "unpack" x, and see if it is a notation for a unary function. If it is, compute the value of that function on input y.

You may already be convinced that (with some work!) one can write a program (say, in Java or C++) that does this; and now we can appeal to the Church-Turing thesis, which says that anything that, intuitively, is computable can be computed by a Turing machine.

Of course, a more direct way to show that $g(x, y)$ is computable is to describe a Turing machine that computes it, explicitly. This would, in particular, avoid the Church-Turing thesis and appeals to intuition. Soon we will have built up enough machinery to show that $g(x, y)$ is computable, appealing to a model of computation that can be *simulated* on a Turing machine: namely, the recursive functions.

2.15 Partial Recursive Functions

To motivate the definition of the recursive functions, note that our proof that there are computable functions that are not primitive recursive actually establishes much more. The argument was simple: all we used was the fact was that it is possible to enumerate functions $f_0, f_1, \ldots$ such that, as a function of x and y, $f_x(y)$ is computable. So the argument applies to *any class of functions that can be enumerated in such a way.* This puts us in a bind: we would like to describe the computable functions explicitly; but any explicit description of a collection of computable functions cannot be exhaustive!

The way out is to allow *partial* functions to come into play. We will see that it *is* possible to enumerate the partial computable functions. In fact, we already pretty much know that this is the case, since it is possible to enumerate Turing machines in a systematic way. We will come back to our diagonal argument later, and explore why it does not go through when partial functions are included.

The question is now this: what do we need to add to the primitive recursive functions to obtain all the partial recursive functions? We need to do two things:

1. Modify our definition of the primitive recursive functions to allow for partial functions as well.

2. *Add* something to the definition, so that some new partial functions are included.

The first is easy. As before, we will start with zero, successor, and projections, and close under composition and primitive recursion. The only difference is that we have to modify the definitions of composition and primitive recursion to allow for the possibility that some of the terms in the definition are not defined. If f and g are partial functions, we will write $f(x) \downarrow$ to mean that f is defined at x, i.e., x is in the domain of f; and $f(x) \uparrow$ to mean the opposite, i.e., that f is not defined at x. We will use $f(x) \simeq g(x)$ to mean that either $f(x)$ and $g(x)$ are both undefined, or they are both defined and equal. We will use these notations for more complicated terms as well. We will adopt the convention that if h and g_0, ..., g_k all are partial functions, then

$$h(g_0(\vec{x}), \dots, g_k(\vec{x}))$$

is defined if and only if each g_i is defined at $\vec{x}$, and h is defined at $g_0(\vec{x})$, ..., $g_k(\vec{x})$. With this understanding, the definitions of composition and primitive recursion for partial functions is just as above, except that we have to replace "$=$" by "$\simeq$".

What we will add to the definition of the primitive recursive functions to obtain partial functions is the ***unbounded search operator.*** If $f(x, \vec{z})$ is any partial function on the natural numbers, define $\mu x\ f(x, \vec{z})$ to be

> the least x such that $f(0, \vec{z}), f(1, \vec{z}), \dots, f(x, \vec{z})$ are all defined, and $f(x, \vec{z}) = 0$, if such an x exists

with the understanding that $\mu x\ f(x, \vec{z})$ is undefined otherwise. This defines $\mu x\ f(x, \vec{z})$ uniquely.

Note that our definition makes no reference to Turing machines, or algorithms, or any specific computational model. But like composition and primitive recursion, there is an operational, computational intuition behind unbounded search. When it comes to the computability of a partial function, arguments where the function is undefined correspond to inputs for which the computation does not halt. The procedure for computing

$\mu x\, f(x, \vec{z})$ will amount to this: compute $f(0, \vec{z}), f(1, \vec{z}), f(2, \vec{z})$ until a value of 0 is returned. If any of the intermediate computations do not halt, however, neither does the computation of $\mu x\, f(x, \vec{z})$.

If $R(x, \vec{z})$ is any relation, $\mu x\, R(x, \vec{z})$ is defined to be $\mu x\, (1 \dot{-} \chi_R(x, \vec{z}))$. In other words, $\mu x\, R(x, \vec{z})$ returns the least value of x such that $R(x, \vec{z})$ holds. So, if $f(x, \vec{z})$ is a total function, $\mu x\, f(x, \vec{z})$ is the same as $\mu x\, (f(x, \vec{z}) = 0)$. But note that our original definition is more general, since it allows for the possibility that $f(x, \vec{z})$ is not everywhere defined (whereas, in contrast, the characteristic function of a relation is always total).

Definition 2.26. The set of *partial recursive functions* is the smallest set of partial functions from the natural numbers to the natural numbers (of various arities) containing zero, successor, and projections, and closed under composition, primitive recursion, and unbounded search.

Of course, some of the partial recursive functions will happen to be total, i.e., defined for every argument.

Definition 2.27. The set of *recursive functions* is the set of partial recursive functions that are total.

A recursive function is sometimes called "total recursive" to emphasize that it is defined everywhere.

2.16 The Normal Form Theorem

Theorem 2.28 (Kleene's Normal Form Theorem). *There is a primitive recursive relation $T(e, x, s)$ and a primitive recursive function $U(s)$, with the following property: if f is any partial recursive function,* *then for some e,*

$$f(x) \simeq U(\mu s\ T(e, x, s))$$

for every x.

The proof of the normal form theorem is involved, but the basic idea is simple. Every partial recursive function has an *index* e, intuitively, a number coding its program or definition. If $f(x) \downarrow$, the computation can be recorded systematically and coded by some number s, and that s codes the computation of f on input x can be checked primitive recursively using only x and the definition e. This means that T is primitive recursive. Given the full record of the computation s, the "upshot" of s is the value of $f(x)$, and it can be obtained from s primitive recursively as well.

The normal form theorem shows that only a single unbounded search is required for the definition of any partial recursive function. We can use the numbers e as "names" of partial recursive functions, and write φ_e for the function f defined by the equation in the theorem. Note that any partial recursive function can have more than one index—in fact, every partial recursive function has infinitely many indices.

2.17 The Halting Problem

The *halting problem* in general is the problem of deciding, given the specification e (e.g., program) of a computable function and a number n, whether the computation of the function on input n halts, i.e., produces a result. Famously, Alan Turing proved that this problem itself cannot be solved by a computable function, i.e., the function

$$h(e, n) = \begin{cases} 1 & \text{if computation } e \text{ halts on input } n \\ 0 & \text{otherwise,} \end{cases}$$

is not computable.

In the context of partial recursive functions, the role of the specification of a program may be played by the index e given in Kleene's normal form theorem. If f is a partial recursive function, any e for which the equation in the normal form theorem holds, is an index of f. Given a number e, the normal form theorem states that

$$\varphi_e(x) \simeq U(\mu s\ T(e, x, s))$$

is partial recursive, and for every partial recursive $f\colon \mathbb{N} \to \mathbb{N}$, there is an $e \in \mathbb{N}$ such that $\varphi_e(x) \simeq f(x)$ for all $x \in \mathbb{N}$. In fact, for each such f there is not just one, but infinitely many such e. The *halting function* h is defined by

$$h(e, x) = \begin{cases} 1 & \text{if } \varphi_e(x) \downarrow \\ 0 & \text{otherwise.} \end{cases}$$

Note that $h(e, x) = 0$ if $\varphi_e(x) \uparrow$, but also when e is not the index of a partial recursive function at all.

Theorem 2.29. *The halting function h is not partial recursive.*

Proof. If h were partial recursive, we could define

$$d(y) = \begin{cases} 1 & \text{if } h(y, y) = 0 \\ \mu x\ x \neq x & \text{otherwise.} \end{cases}$$

From this definition it follows that

1. $d(y) \downarrow$ iff $\varphi_y(y) \uparrow$ or y is not the index of a partial recursive function.

2. $d(y) \uparrow$ iff $\varphi_y(y) \downarrow$.

If h were partial recursive, then d would be partial recursive as well. Thus, by the Kleene normal form theorem, it has an index e_d. Consider the value of $h(e_d, e_d)$. There are two possible cases, 0 and 1.

1. If $h(e_d, e_d) = 1$ then $\varphi_{e_d}(e_d) \downarrow$. But $\varphi_{e_d} \simeq d$, and $d(e_d)$ is defined iff $h(e_d, e_d) = 0$. So $h(e_d, e_d) \neq 1$.

2. If $h(e_d, e_d) = 0$ then either e_d is not the index of a partial recursive function, or it is and $\varphi_{e_d}(e_d) \uparrow$. But again, $\varphi_{e_d} \simeq d$, and $d(e_d)$ is undefined iff $\varphi_{e_d}(e_d) \downarrow$.

The upshot is that e_d cannot, after all, be the index of a partial recursive function. But if h were partial recursive, d would be too, and so our definition of e_d as an index of it would be admissible. We must conclude that h cannot be partial recursive. □

2.18 General Recursive Functions

There is another way to obtain a set of total functions. Say a total function $f(x, \vec{z})$ is *regular* if for every sequence of natural numbers $\vec{z}$, there is an x such that $f(x, \vec{z}) = 0$. In other words, the regular functions are exactly those functions to which one can apply unbounded search, and end up with a total function. One can, conservatively, restrict unbounded search to regular functions:

Definition 2.30. The set of *general recursive functions* is the smallest set of functions from the natural numbers to the natural numbers (of various arities) containing zero, successor, and projections, and closed under composition, primitive recursion, and unbounded search applied to *regular* functions.

Clearly every general recursive function is total. The difference between Definition 2.30 and Definition 2.27 is that in the latter one is allowed to use partial recursive functions along the way; the only requirement is that the function you end up with at the end is total. So the word "general," a historic relic, is a misnomer; on the surface, Definition 2.30 is *less* general than Definition 2.27. But, fortunately, the difference is illusory; though the definitions are different, the set of general recursive functions and the set of recursive functions are one and the same.

Summary

In order to show that **Q** represents all computable functions, we need a precise model of computability that we can take as the basis for a proof. There are, of course, many models of computability, such as Turing machines. One model that plays a significant role historically—it's one of the first models proposed, and is also the one used by Gödel himself—is that of the **recursive functions**. The recursive functions are a class of arithmetical functions—that is, their domain and range are the natural numbers—that can be defined from a few basic functions using a few operations. The basic functions are zero, succ, and the projection functions. The operations are **composition**, **primitive recursion**, and **regular minimization**. Composition is simply a general version of "chaining together" functions: first apply one, then apply the other to the result. Primitive recursion defines a new function f from two functions g, h already defined, by stipulating that the value of f for 0 is given by g, and the value for any number $n + 1$ is given by h applied to $f(n)$. Functions that can be defined using just these two principles are called **primitive recursive**. A relation is primitive recursive iff its characteristic function is. It turns out that a whole list of interesting functions and relations is primitive recursive (such as addition, multiplication, exponentiation, divisibility), and that we can define new primitive recursive functions and relations from old ones using principles such as bounded quantification and bounded minimization. In particular, this allowed us to show that we can deal with **sequences** of numbers in primitive recursive ways. That is, there is a way to "code" sequences of numbers as single numbers in such a way that we can compute the i-the element, the length, the concatenation of two sequences, etc., all using primitive recursive functions operating on these codes. To obtain all the computable functions, we finally added definition by **regular minimization** to composition and primitive recursion. A function $g(x, y)$ is **regular** iff, for every y it takes the value 0 for at last one x. If f is regular, the least x such that $g(x, y) = 0$ always exists, and can be found simply by computing all the values of $g(0,y)$, $g(1,y)$, etc., until one of them is $= 0$. The resulting function $f(y) = \mu x\ g(x,y) = 0$ is the function defined by regular minimization from g. It is always total and computable. The resulting set of functions are called **general recursive**. One version of the Church-Turing Thesis says that the computable arithmetical functions are exactly the general recursive ones.

Problems

Problem 2.1. Prove Proposition 2.5 by showing that the primitive recursive definition of mult is can be put into the form required by Definition 2.1 and showing that the corresponding functions f and g are primitive recursive.

Problem 2.2. Give the complete primitive recursive notation for mult.

Problem 2.3. Prove Proposition 2.13.

Problem 2.4. Show that

$$f(x,y) = 2^{(2^{\cdot^{\cdot^{\cdot^{2^x}}}})}\Big\}y \text{ 2's}$$

is primitive recursive.

Problem 2.5. Show that integer division $d(x,y) = \lfloor x/y \rfloor$ (i.e., division, where you disregard everything after the decimal point) is primitive recursive. When $y = 0$, we stipulate $d(x,y) = 0$. Give an explicit definition of d using primitive recursion and composition.

Problem 2.6. Suppose $R(\vec{x}, z)$ is primitive recursive. Define the function $m'_R(\vec{x}, y)$ which returns the least z less than y such that $R(\vec{x}, z)$ holds, if there is one, and 0 otherwise, by primitive recursion from χ_R.

Problem 2.7. Define integer division $d(x, y)$ using bounded minimization.

Problem 2.8. Show that there is a primitive recursive function sconcat(s) with the property that

$$\text{sconcat}(\langle s_0, \ldots, s_k \rangle) = s_0 \frown \ldots \frown s_k.$$

Problem 2.9. Show that there is a primitive recursive function tail(s) with the property that

$$\text{tail}(\Lambda) = 0 \text{ and}$$
$$\text{tail}(\langle s_0, \ldots, s_k \rangle) = \langle s_1, \ldots, s_k \rangle.$$

Problem 2.10. Prove Proposition 2.24.

Problem 2.11. The definition of hSubtreeSeq in the proof of Proposition 2.25 in general includes repetitions. Give an alternative definition which guarantees that the code of a subtree occurs only once in the resulting list.

CHAPTER 3

Arithmetization of Syntax

3.1 Introduction

In order to connect computability and logic, we need a way to talk about the objects of logic (symbols, terms, formulas, derivations), operations on them, and their properties and relations, in a way amenable to computational treatment. We can do this directly, by considering computable functions and relations on symbols, sequences of symbols, and other objects built from them. Since the objects of logical syntax are all finite and built from a countable sets of symbols, this is possible for some models of computation. But other models of computation—such as the recursive functions—are restricted to numbers, their relations and functions. Moreover, ultimately we also want to be able to deal with syntax within certain theories, specifically, in theories formulated in the language of arithmetic. In these cases it is necessary to *arithmetize* syntax, i.e., to represent syntactic objects, operations on them, and their relations, as numbers, arithmetical functions, and arithmetical relations, respectively. The idea, which goes back to Leibniz, is to assign numbers to syntactic objects.

It is relatively straightforward to assign numbers to symbols as their "codes." Some symbols pose a bit of a challenge, since, e.g., there are infinitely many variables, and even infinitely many function symbols of each arity n. But of course it's possible to assign numbers to symbols systematically in such a way that, say, v_2 and v_3 are assigned different codes. Sequences of symbols (such as terms and formulas) are a bigger challenge. But if can deal with sequences of numbers purely arithmetically (e.g., by the powers-of-primes coding of sequences), we can extend the coding of individual symbols to coding of sequences of symbols, and then further to sequences or other arrangements of formulas, such as derivations. This extended coding is called "Gödel numbering." Every term, formula, and derivation is assigned a Gödel number.

By coding sequences of symbols as sequences of their codes, and by chosing a system of coding sequences that can be dealt with using computable functions, we can then also deal with Gödel numbers using computable functions. In practice, all the relevant functions will be primitive recursive. For instance, computing the length of a sequence and computing the i-th element of a sequence from the code of the sequence are both primitive recursive. If the number coding the sequence is, e.g., the Gödel number of a formula A, we immediately see that the length of a formula and the (code of the) i-th symbol in a formula can also be computed from the Gödel number of A. It is a bit harder to prove that, e.g., the property of being the Gödel number of a correctly formed term, of being the Gödel number of a corret derivation is primitive recursive. It is nevertheless possible, because the sequences of interest (terms, formulas, derivations) are inductively defined.

As an example, consider the operation of substitution. If A is a formula, x a variable, and t a term, then $A[t/x]$ is the result of replacing every free occurrence of x in A by t. Now suppose we have assigned Gödel numbers to A, x, t—say, k, l, and m, respectively. The same scheme assigns a Gödel number to $A[t/x]$, say, n. This mapping—of k, l, and m to n—is the arithmetical analog of the substitution operation. When the substitution operation maps A, x, t to $A[t/x]$, the arithmetized substitution functions maps the Gödel numbers k, l, m to the Gödel number n. We will see that this function is primitive recursive.

Arithmetization of syntax is not just of abstract interest, although it was originally a non-trivial insight that languages like the language of arithmetic, which do not come with mechanisms for "talking about" languages can, after all, formalize complex properties of expressions. It is then just a small step to ask what a theory in this language, such as Peano arithmetic, can *prove* about its own language (including, e.g., whether sentences are provable or true). This leads us to the famous limitative theorems of Gödel (about unprovability) and Tarski (the undefinability of truth). But the trick of arithmetizing syntax is also important in order to prove some important results in computability theory, e.g., about the computational prower of theories or the relationship between different models of computability. The arithmetization of syntax serves as a model for arithmetizing other objects and properties. For instance, it is similarly possible to arithmetize configurations and computations (say, of Turing machines). This makes it possible to simulate computations in one model (e.g., Turing machines) in another (e.g., recursive functions).

3.2 Coding Symbols

The basic language $\mathcal{L}$ of first order logic makes use of the symbols

$$\bot \quad \neg \quad \lor \quad \land \quad \rightarrow \quad \forall \quad \exists \quad = \quad (\quad) \quad ,$$

together with countable sets of variables and constant symbols, and countable sets of function symbols and predicate symbols of arbitrary arity. We can assign *codes* to each of these symbols in such a way that every symbol is assigned a unique number as its code, and no two different symbols are assigned the same number. We know that this is possible since the set of all symbols is countable and so there is a bijection between it and the set of natural numbers. But we want to make sure that we can recover the symbol (as well as some information about it, e.g., the arity of a function symbol) from its code in a computable way. There are many possible ways of doing this, of course. Here is one such way, which uses primitive recursive functions. (Recall that $\langle n_0, \ldots, n_k \rangle$ is the number coding the sequence of numbers $n_0, \ldots, n_k$.)

Definition 3.1. If s is a symbol of $\mathcal{L}$, let the *symbol code* c_s be defined as follows:

1. If s is among the logical symbols, c_s is given by the following table:

$\bot$	$\neg$	$\lor$	$\land$	$\rightarrow$	$\forall$
$\langle 0,0 \rangle$	$\langle 0,1 \rangle$	$\langle 0,2 \rangle$	$\langle 0,3 \rangle$	$\langle 0,4 \rangle$	$\langle 0,5 \rangle$
$\exists$	$=$	(	)	,	
$\langle 0,6 \rangle$	$\langle 0,7 \rangle$	$\langle 0,8 \rangle$	$\langle 0,9 \rangle$	$\langle 0,10 \rangle$	

2. If s is the i-th variable v_i, then $c_s = \langle 1, i \rangle$.

3. If s is the i-th constant symbol c_i^n, then $c_s = \langle 2, i \rangle$.

4. If s is the i-th n-ary function symbol f_i^n, then $c_s = \langle 3, n, i \rangle$.

5. If s is the i-th n-ary predicate symbol P_i^n, then $c_s = \langle 4, n, i \rangle$.

Proposition 3.2. *The following relations are primitive recursive:*

1. $\mathrm{Fn}(x, n)$ *iff x is the code of f_i^n for some i, i.e., x is the code of an n-ary function symbol.*

2. $\mathrm{Pred}(x, n)$ *iff x is the code of P_i^n for some i or x is the code of $=$ and $n = 2$, i.e., x is the code of an n-ary predicate symbol.*

Definition 3.3. If $s_0, \dots, s_{n-1}$ is a sequence of symbols, its *Gödel number* is $\langle c_{s_0}, \dots, c_{s_{n-1}} \rangle$.

Note that *codes* and *Gödel numbers* are different things. For instance, the variable v_5 has a code $c_{v_5} = \langle 1, 5 \rangle = 2^2 \cdot 3^6$. But the variable v_5 considered as a term is also a sequence of symbols (of length 1). The *Gödel number* ${}^{\#}v_5{}^{\#}$ of the *term* v_5 is $\langle c_{v_5} \rangle = 2^{c_{v_5}+1} = 2^{2^2 \cdot 3^6 + 1}$.

Example 3.4. Recall that if $k_0, \dots, k_{n-1}$ is a sequence of numbers, then the code of the sequence $\langle k_0, \dots, k_{n-1} \rangle$ in the power-of-primes coding is

$$2^{k_0+1} \cdot 3^{k_1+1} \cdot \dots \cdot p_{n-1}^{k_{n-1}},$$

where p_i is the i-th prime (starting with $p_0 = 2$). So for instance, the formula $v_0 = 0$, or, more explicitly, $=(v_0, c_0)$, has the Gödel number

$$\langle c_{=}, c_{(}, c_{v_0}, c_{,}, c_{c_0}, c_{)} \rangle.$$

Here, $c_{=}$ is $\langle 0, 7 \rangle = 2^{0+1} \cdot 3^{7=1}$, c_{v_0} is $\langle 1, 0 \rangle = 2^{1+1} \cdot 3^{0+1}$, etc. So ${}^{\#}{=}(v_0, c_0)^{\#}$ is

$$\begin{aligned} 2^{c_{=}+1} \cdot 3^{c_{(}+1} \cdot 5^{c_{v_0}+1} \cdot 7^{c_{,}+1} \cdot 11^{c_{c_0}+1} \cdot 13^{c_{)}+1} &= \\ 2^{2^1 \cdot 3^8+1} \cdot 3^{2^1 \cdot 3^9+1} \cdot 5^{2^2 \cdot 3^1+1} \cdot 7^{2^1 \cdot 3^{11}+1} \cdot 11^{2^3 \cdot 3^1+1} \cdot 13^{2^1 \cdot 3^{10}+1} &= \\ 2^{13\,123} \cdot 3^{39\,367} \cdot 5^{13} \cdot 7^{354\,295} \cdot 11^{25} \cdot 13^{118\,099}&. \end{aligned}$$

3.3 Coding Terms

A term is simply a certain kind of sequence of symbols: it is built up inductively from constants and variables according to the formation rules for terms. Since sequences of symbols can be coded as numbers—using a coding scheme for the symbols plus a way to code sequences of numbers—assigning Gödel numbers to terms is not difficult. The challenge is rather to show that the property a number has if it is the Gödel number of a correctly formed term is computable, or in fact primitive recursive.

Variables and constant symbols are the simplest terms, and testing whether x is the Gödel number of such a term is easy: $\mathrm{Var}(x)$ holds if x is ${}^{\#}v_i{}^{\#}$ for some i. In other words, x is a sequence of length 1 and its single element $(x)_0$ is the code of some variable v_i, i.e., x is $\langle\langle 1, i\rangle\rangle$ for some i. Similarly, $\mathrm{Const}(x)$ holds if x is ${}^{\#}c_i{}^{\#}$ for some i. Both of these relations are primitive recursive, since if such an i exists, it must be $< x$:

$$\mathrm{Var}(x) \Leftrightarrow (\exists i < x)\; x = \langle\langle 1, i\rangle\rangle$$
$$\mathrm{Const}(x) \Leftrightarrow (\exists i < x)\; x = \langle\langle 2, i\rangle\rangle$$

Proposition 3.5. *The relations* $\mathrm{Term}(x)$ *and* $\mathrm{ClTerm}(x)$ *which hold iff x is the Gödel number of a term or a closed term, respectively, are primitive recursive.*

Proof. A sequence of symbols s is a term iff there is a sequence s_0, …, $s_{k-1} = s$ of terms which records how the term s was formed from constant symbols and variables according to the formation rules for terms. To express that such a putative formation sequence follows the formation rules it has to be the case that, for each $i < k$, either

1. s_i is a variable v_j, or
2. s_i is a constant symbol c_j, or
3. s_i is built from n terms t_1, …, t_n occurring prior to place i using an n-place function symbol f^n_j.

To show that the corresponding relation on Gödel numbers is primitive recursive, we have to express this condition primitive recursively, i.e., using primitive recursive functions, relations, and bounded quantification.

Suppose y is the number that codes the sequence s_0, …, s_{k-1}, i.e., $y = \langle {}^{\#}s_0{}^{\#}, \dots, {}^{\#}s_k{}^{\#}\rangle$. It codes a formation sequence for the term with Gödel number x iff for all $i < k$:

1. $\mathrm{Var}((y)_i)$, or

2. $\mathrm{Const}((y)_i)$, or

3. there is an n and a number $z = \langle z_1, \dots, z_n \rangle$ such that each z_l is equal to some $(y)_{i'}$ for $i' < i$ and

$$(y)_i = {}^{\#}f^n_j({}^{\#} \frown \mathrm{flatten}(z) \frown {}^{\#}){}^{\#},$$

and moreover $(y)_{k-1} = x$. (The function $\mathrm{flatten}(z)$ turns the sequence $\langle {}^{\#}t_1{}^{\#}, \dots, {}^{\#}t_n{}^{\#} \rangle$ into ${}^{\#}t_1, \dots, t_n{}^{\#}$ and is primitive recursive.)

The indices j, n, the Gödel numbers z_l of the terms t_l, and the code z of the sequence $\langle z_1, \dots, z_n \rangle$, in (3) are all less than y. We can replace k above with $\mathrm{len}(y)$. Hence we can express "y is the code of a formation sequence of the term with Gödel number x" in a way that shows that this relation is primitive recursive.

We now just have to convince ourselves that there is a primitive recursive bound on y. But if x is the Gödel number of a term, it must have a formation sequence with at most $\mathrm{len}(x)$ terms (since every term in the formation sequence of s must start at some place in s, and no two subterms can start at the same place). The Gödel number of each subterm of s is of course $\leq x$. Hence, there always is a formation sequence with code $\leq x^{\mathrm{len}(x)}$.

For ClTerm, simply leave out the clause for variables. □

Proposition 3.6. *The function* $\mathrm{num}(n) = {}^{\#}\overline{n}{}^{\#}$ *is primitive recursive.*

Proof. We define $\mathrm{num}(n)$ by primitive recursion:

$$\begin{aligned} \mathrm{num}(0) &= {}^{\#}\mathrm{o}{}^{\#} \\ \mathrm{num}(n+1) &= {}^{\#}\prime({}^{\#} \frown \mathrm{num}(n) \frown {}^{\#}){}^{\#}. \end{aligned}$$

□

3.4 Coding Formulas

Proposition 3.7. *The relation* Atom(x) *which holds iff x is the Gödel number of an atomic formula, is primitive recursive.*

Proof. The number x is the Gödel number of an atomic formula iff one of the following holds:

1. There are n, $j < x$, and $z < x$ such that for each $i < n$, $\mathrm{Term}((z)_i)$ and $x =$

$$^{\#}P^n_j(^{\#} \frown \mathrm{flatten}(z) \frown {}^{\#})^{\#}.$$

2. There are $z_1, z_2 < x$ such that $\mathrm{Term}(z_1)$, $\mathrm{Term}(z_2)$, and $x =$

$$^{\#}{=}(^{\#} \frown z_1 \frown {}^{\#},^{\#} \frown z_2 \frown {}^{\#})^{\#}.$$

3. $x = {}^{\#}\bot^{\#}$. □

Proposition 3.8. *The relation* Frm(x) *which holds iff x is the Gödel number of a formula is primitive recursive.*

Proof. A sequence of symbols s is a formula iff there is formation sequence s_0, ..., $s_{k-1} = s$ of formula which records how s was formed from atomic formulas according to the formation rules. The code for each s_i (and indeed of the code of the sequence $\langle s_0, \ldots, s_{k-1}\rangle$ is less than the code x of s. □

Proposition 3.9. *The relation* FreeOcc(x, z, i), *which holds iff the i-th symbol of the formula with Gödel number x is a free occurrence of the variable with Gödel number z, is primitive recursive.*

Proof. Exercise. □

Proposition 3.10. *The property* Sent(x) *which holds iff x is the Gödel number of a sentence is primitive recursive.*

Proof. A sentence is a formula without free occurrences of variables. So Sent(x) holds iff

$$(\forall i < \mathrm{len}(x))\,(\forall z < x)$$
$$((\exists j < z)\; z = {}^{\#}v_j{}^{\#} \to \neg\mathrm{FreeOcc}(x, z, i)). \quad \square$$

3.5 Substitution

Recall that substitution is the operation of replacing all free occurrences of a variable u in a formula A by a term t, written $A[t/u]$. This operation, when carried out on Gödel numbers of variables, formulas, and terms, is primitive recursive.

Proposition 3.11. *There is a primitive recursive function* Subst(x, y, z) *with the property that*

$$\mathrm{Subst}({}^{\#}A^{\#}, {}^{\#}t^{\#}, {}^{\#}u^{\#}) = {}^{\#}A[t/u]^{\#}$$

Proof. We can then define a function hSubst by primitive recursion as follows:

$$\mathrm{hSubst}(x, y, z, 0) = \Lambda$$
$$\mathrm{hSubst}(x, y, z, i+1) =$$
$$\begin{cases} \mathrm{hSubst}(x, y, z, i) \frown y & \text{if } \mathrm{FreeOcc}(x, z, i) \\ \mathrm{append}(\mathrm{hSubst}(x, y, z, i), (x)_i) & \text{otherwise.} \end{cases}$$

Subst(x, y, z) can now be defined as hSubst(x, y, z, len(x)). $\square$

Proposition 3.12. *The relation* FreeFor(x, y, z), *which holds iff the term with Gödel number y is free for the variable with Gödel number z*

in the formula with Gödel number x, is primitive recursive.

Proof. Exercise. □

3.6 Derivations in Natural Deduction

In order to arithmetize derivations, we must represent derivations as numbers. Since derivations are trees of formulas where each inference carries one or two labels, a recursive representation is the most obvious approach: we represent a derivation as a tuple, the components of which are the number of immediate sub-derivations leading to the premises of the last inference, the representations of these sub-derivations, and the end-formula, the discharge label of the last inference, and a number indicating the type of the last inference.

Definition 3.13. If δ is a derivation in natural deduction, then $^{\#}\delta^{\#}$ is defined inductively as follows:

1. If δ consists only of the assumption A, then $^{\#}\delta^{\#}$ is $\langle 0, {}^{\#}A^{\#}, n \rangle$. The number n is 0 if it is an undischarged assumption, and the numerical label otherwise.

2. If δ ends in an inference with one, two, or three premises, then $^{\#}\delta^{\#}$ is

$$\begin{aligned} &\langle 1, {}^{\#}\delta_1{}^{\#}, {}^{\#}A^{\#}, n, k \rangle, \\ &\langle 2, {}^{\#}\delta_1{}^{\#}, {}^{\#}\delta_2{}^{\#}, {}^{\#}A^{\#}, n, k \rangle, \text{ or} \\ &\langle 3, {}^{\#}\delta_1{}^{\#}, {}^{\#}\delta_2{}^{\#}, {}^{\#}\delta_3{}^{\#}, {}^{\#}A^{\#}, n, k \rangle, \end{aligned}$$

respectively. Here δ_1, δ_2, δ_3 are the sub-derivations ending in the premise(s) of the last inference in δ, A is the conclusion of the last inference in δ, n is the discharge label of the last inference (0 if the inference does not discharge any assumptions), and k is given by the following table according to which rule was used in the last inference.

Rule:	$\wedge$Intro	$\wedge$Elim	$\vee$Intro	$\vee$Elim
k:	1	2	3	4
Rule:	$\to$Intro	$\to$Elim	$\neg$Intro	$\neg$Elim
k:	5	6	7	8
Rule:	$\bot_I$	$\bot_C$	$\forall$Intro	$\forall$Elim
k:	9	10	11	12
Rule:	$\exists$Intro	$\exists$Elim	$=$Intro	$=$Elim
k:	13	14	15	16

Example 3.14. Consider the very simple derivation

$$1\,\frac{\dfrac{[A \wedge B]^1}{A}\,\wedge\text{Elim}}{(A \wedge B) \to A}\,\to\text{Intro}$$

The Gödel number of the assumption would be $d_0 = \langle 0, {}^{\#}A \wedge B^{\#}, 1\rangle$. The Gödel number of the derivation ending in the conclusion of $\wedge$Elim would be $d_1 = \langle 1, d_0, {}^{\#}A^{\#}, 0, 2\rangle$ (1 since $\wedge$Elim has one premise, the Gödel number of conclusion A, 0 because no assumption is discharged, and 2 is the number coding $\wedge$Elim). The Gödel number of the entire derivation then is $\langle 1, d_1, {}^{\#}((A \wedge B) \to A)^{\#}, 1, 5\rangle$, i.e.,

$$\langle 1, \langle 1, \langle 0, {}^{\#}(A \wedge B)^{\#}, 1\rangle, {}^{\#}A^{\#}, 0, 2\rangle, {}^{\#}((A \wedge B) \to A)^{\#}, 1, 5\rangle.$$

Having settled on a representation of derivations, we must also show that we can manipulate Gödel numbers of such derivations primitive recursively, and express their essential properties and relations. Some operations are simple: e.g., given a Gödel number d of a derivation, $\text{EndFmla}(d) = (d)_{(d)_0+1}$ gives us the Gödel number of its end-formula, $\text{DischargeLabel}(d) = (d)_{(d)_0+2}$ gives us the discharge label and $\text{LastRule}(d) = (d)_{(d)_0+3}$ the number indicating the type of the last inference. Some are much harder. We'll at least sketch how to do this. The goal is to show that the relation "δ is a derivation of A from Γ" is a primitive recursive relation of the Gödel numbers of δ and A.

Proposition 3.15. *The following relations are primitive recursive:*

1. *A occurs as an assumption in δ with label n.*

2. *All assumptions in δ with label n are of the form A (i.e., we can discharge the assumption A using label n in δ).*

Proof. We have to show that the corresponding relations between Gödel numbers of formulas and Gödel numbers of derivations are primitive recursive.

1. We want to show that $\mathrm{Assum}(x, d, n)$, which holds if x is the Gödel number of an assumption of the derivation with Gödel number d labelled n, is primitive recursive. This is the case if the derivation with Gödel number $\langle 0, x, n\rangle$ is a sub-derivation of d. Note that the way we code derivations is a special case of the coding of trees introduced in section 2.12, so the primitive recursive function $\mathrm{SubtreeSeq}(d)$ gives a sequence of Gödel numbers of all sub-derivations of d (of length a most d). So we can define

$$\mathrm{Assum}(x, d, n) \Leftrightarrow (\exists i < d)\, (\mathrm{SubtreeSeq}(d))_i = \langle 0, x, n\rangle.$$

2. We want to show that $\mathrm{Discharge}(x, d, n)$, which holds if all assumptions with label n in the derivation with Gödel number d all are the formula with Gödel number x. But this relation holds iff $(\forall y < d)\, (\mathrm{Assum}(y, d, n) \rightarrow y = x)$. □

Proposition 3.16. *The property* Correct(d) *which holds iff the last inference in the derivation δ with Gödel number d is correct, is primitive recursive.*

Proof. Here we have to show that for each rule of inference R the relation $\text{FollowsBy}_R(d)$ is primitive recursive, where $\text{FollowsBy}_R(d)$ holds iff d is the Gödel number of derivation δ, and the end-formula of δ follows by a correct application of R from the immediate sub-derivations of δ.

A simple case is that of the $\wedge$Intro rule. If δ ends in a correct $\wedge$Intro inference, it looks like this:

$$\dfrac{\begin{matrix}\vdots\,\delta_1 & \vdots\,\delta_2 \\ A & B\end{matrix}}{A \wedge B}\;\wedge\text{Intro}$$

Then the Gödel number d of δ is $\langle 2, d_1, d_2, {}^{\#}(A \wedge B)^{\#}, 0, k\rangle$ where $\text{EndFmla}(d_1) = {}^{\#}A^{\#}$, $\text{EndFmla}(d_2) = {}^{\#}B^{\#}$, $n = 0$, and $k = 1$. So we can define $\text{FollowsBy}_{\wedge\text{Intro}}(d)$ as

$$(d)_0 = 2 \wedge \text{DischargeLabel}(d) = 0 \wedge \text{LastRule}(d) = 1 \wedge$$
$$\text{EndFmla}(d) = {}^{\#}({}^{\#} \frown \text{EndFmla}((d)_1) \frown {}^{\#}\wedge^{\#} \frown \text{EndFmla}((d)_2) \frown {}^{\#})^{\#}.$$

Another simple example if the =Intro rule. Here the premise is an empty derivation, i.e., $(d)_1 = 0$, and no discharge label, i.e., $n = 0$. However, A must be of the form $t = t$, for a closed term t. Here, a primitive recursive definition is

$$(d)_0 = 1 \wedge (d)_1 = 0 \wedge \text{DischargeLabel}(d) = 0 \wedge$$
$$(\exists t < d)\;(\text{ClTerm}(t) \wedge \text{EndFmla}(d) = {}^{\#}{=}({}^{\#} \frown t \frown {}^{\#},{}^{\#} \frown t \frown {}^{\#}){}^{\#})$$

For a more complicated example, $\text{FollowsBy}_{\rightarrow\text{Intro}}(d)$ holds iff the end-formula of δ is of the form $(A \rightarrow B)$, where the end-formula of δ_1 is B, and any assumption in δ labelled n is of the form A. We can express this primitive recursively by

$$
\begin{aligned}
&(d)_0 = 1 \land \\
&\qquad (\exists a < d)\,(\text{Discharge}(a, (d)_1, \text{DischargeLabel}(d)) \land \\
&\quad \text{EndFmla}(d) = ({}^{\#}({}^{\#} \frown a \frown {}^{\#}{\to}^{\#} \frown \text{EndFmla}((d)_1) \frown {}^{\#}){}^{\#}))
\end{aligned}
$$

(Think of a as the Gödel number of A).

For another example, consider $\exists$Intro. Here, the last inference in δ is correct iff there is a formula A, a closed term t and a variable x such that $A[t/x]$ is the end-formula of the derivation δ_1 and $\exists x\, A$ is the conclusion of the last inference. So, $\text{FollowsBy}_{\exists\text{Intro}}(d)$ holds iff

$$
\begin{aligned}
&(d)_0 = 1 \land \text{DischargeLabel}(d) = 0 \land \\
&\qquad (\exists a < d)\,(\exists x < d)\,(\exists t < d)\,(\text{ClTerm}(t) \land \text{Var}(x) \land \\
&\text{Subst}(a, t, x) = \text{EndFmla}((d)_1) \land \text{EndFmla}(d) = ({}^{\#}\exists^{\#} \frown x \frown a)).
\end{aligned}
$$

We then define $\text{Correct}(d)$ as

$$
\begin{aligned}
&\text{Sent}(\text{EndFmla}(d)) \land \\
&\qquad (\text{LastRule}(d) = 1 \land \text{FollowsBy}_{\land\text{Intro}}(d)) \lor \cdots \lor \\
&\qquad\quad (\text{LastRule}(d) = 16 \land \text{FollowsBy}_{=\text{Elim}}(d)) \lor \\
&\qquad\qquad\qquad (\exists n < d)\,(\exists x < d)\,(d = \langle 0, x, n \rangle).
\end{aligned}
$$

The first line ensures that the end-formula of d is a sentence. The last line covers the case where d is just an assumption. □

Proposition 3.17. *The relation* $\text{Deriv}(d)$ *which holds if* d *is the Gödel number of a correct derivation* δ*, is primitive recursive.*

Proof. A derivation δ is correct if every one of its inferences is a correct application of a rule, i.e., if every one of its sub-derivations ends in a correct inference. So, $\text{Deriv}(d)$ iff

$$(\forall i < \text{len}(\text{SubtreeSeq}(d)))\ \text{Correct}((\text{SubtreeSeq}(d))_i)$$ □

Proposition 3.18. *The relation* OpenAssum(z, d) *that holds if z is the Gödel number of an undischarged assumption A of the derivation δ with Gödel number d, is primitive recursive.*

Proof. An occurrence of an assumption is discharged if it occurs with label n in a sub-derivation of δ that ends in a rule with discharge label n. So A is an undischarged assumption of δ if at least one of its occurrences is not discharged in δ. We must be careful: δ may contain both discharged and undischarged occurrences of A.

Consider a sequence $\delta_0, \dots, \delta_k$ where $\delta_0 = d$, δ_k is the assumption $[A]^n$ (for some n), and δ_i is an immediate sub-derivation of δ_{i+1}. If such a sequence exists in which no δ_i ends in an inference with discharge label n, then A is an undischarged assumption of δ.

The primitive recursive function SubtreeSeq(d) provides us with a sequence of Gödel numbers of all sub-derivations of δ. Any sequence of Gödel numbers of sub-derivations of δ is a subsequence of it. Being a subsequence of is a primitive recursive relation: Subseq(s, s') holds iff $(\forall i < \mathrm{len}(s))\, \exists j < \mathrm{len}(s')\, (s)_i = (s)_j$. Being an immediate sub-derivation is as well: Subderiv(d, d') iff $(\exists j < (d')_0)\; d = (d')_j$. So we can define OpenAssum(z, d) by

$$(\exists s < \mathrm{SubtreeSeq}(d))\,(\mathrm{Subseq}(s, \mathrm{SubtreeSeq}(d)) \land (s)_0 = d \land$$
$$(\exists n < d)\,((s)_{\mathrm{len}(s) \dot{-} 1} = \langle 0, z, n\rangle \land$$
$$(\forall i < (\mathrm{len}(s) \dot{-} 1))\,(\mathrm{Subderiv}((s)_i, (s)_{i+1})] \land$$
$$\mathrm{DischargeLabel}((s)_{i+1}) \neq n))). \quad \square$$

Proposition 3.19. *Suppose Γ is a primitive recursive set of sentences. Then the relation* Prf$_\Gamma(x, y)$ *expressing "x is the code of a derivation δ of A from undischarged assumptions in Γ and y is the Gödel number of A" is primitive recursive.*

Proof. Suppose "$y \in \Gamma$" is given by the primitive recursive predicate $R_\Gamma(y)$. We have to show that Prf$_\Gamma(x, y)$ which holds iff y is the Gödel number of a sentence A and x is the code of a natural deduction derivation with end formula A and all undischarged assumptions in Γ is primitive recursive.

By Proposition 3.17, the property $\text{Deriv}(x)$ which holds iff x is the Gödel number of a correct derivation δ in natural deduction is primitive recursive. Thus we can define $\text{Prf}_\Gamma(x, y)$ by

$$\begin{aligned}\text{Prf}_\Gamma(x, y) \Leftrightarrow {} & \text{Deriv}(x) \land \text{EndFmla}(x) = y \land \\ & (\forall z < x)\,(\text{OpenAssum}(z, x) \rightarrow R_\Gamma(z)).\end{aligned}$$ □

Summary

The proof of the incompleteness theorems requires that we have a way to talk about provability in a theory (such as **PA**) in the language of the theory itself, i.e., in the language of arithmetic. But the language of arithmetic only deals with numbers, not with formulas or derivations. The solution to this problem is to define a systematic mapping from formulas and derivations to numbers. The number associated with a formula or a derivation is called its **Gödel number**. If A is a formula, $^{\#}A^{\#}$ is its Gödel number. We showed that important operations on formulas turn into primitive recursive functions on the respective Gödel numbers. For instance, $A[t/x]$, the operation of substituting a term t for every free occurrence of x in A, corresponds to an arithmetical function $\text{subst}(n, m, k)$ which, if applied to the Gödel numbers of A, t, and x, yields the Gödel number of $A[t/x]$. In other words, $\text{subst}(^{\#}A^{\#}, {}^{\#}t^{\#}, {}^{\#}x^{\#}) = {}^{\#}A[t/x]^{\#}$. Likewise, properties of derivations turn into primitive recursive relations on the respective Gödel numbers. In particular, the property $\text{Deriv}(n)$ that holds of n if it is the Gödel number of a correct derivation in natural deduction, is primitive recursive. Showing that these are primitive recursive required a fair amount of work, and at times some ingenuity, and depended essentially on the fact that operating with sequences is primitive recursive. If a theory $\mathbf{T}$ is decidable, then we can use Deriv to define a decidable relation $\text{Prf}_{\mathbf{T}}(n, m)$ which holds if n

is the Gödel number of a derivation of the sentence with Gödel number m from $\mathbf{T}$. This relation is primitive recursive if the set of axioms of $\mathbf{T}$ is, and merely general recursive if the axioms of $\mathbf{T}$ are decidable but not primitive recursive.

Problems

Problem 3.1. Show that the function flatten(z), which turns the sequence $\langle {}^{\#}t_1{}^{\#}, \dots, {}^{\#}t_n{}^{\#} \rangle$ into ${}^{\#}t_1, \dots, t_n{}^{\#}$, is primitive recursive.

Problem 3.2. Give a detailed proof of Proposition 3.8 along the lines of the first proof of Proposition 3.5

Problem 3.3. Give a detailed proof of Proposition 3.8 along the lines of the alternate proof of Proposition 3.5

Problem 3.4. Prove Proposition 3.9. You may make use of the fact that any substring of a formula which is a formula is a sub-formula of it.

Problem 3.5. Prove Proposition 3.12

Problem 3.6. Define the following properties as in Proposition 3.16:

1. $\mathrm{FollowsBy}_{\to\mathrm{Elim}}(d)$,
2. $\mathrm{FollowsBy}_{=\mathrm{Elim}}(d)$,
3. $\mathrm{FollowsBy}_{\lor\mathrm{Elim}}(d)$,
4. $\mathrm{FollowsBy}_{\forall\mathrm{Intro}}(d)$.

For the last one, you will have to also show that you can test primitive recursively if the last inference of the derivation with Gödel number d satisfies the eigenvariable condition, i.e., the eigenvariable a of the $\forall$Intro inference occurs neither in the end-formula of d nor in an open assumption of d. You may use the primitive recursive predicate OpenAssum from Proposition 3.18 for this.

CHAPTER 4

Representability in Q

4.1 Introduction

The incompleteness theorems apply to theories in which basic facts about computable functions can be expressed and proved. We will describe a very minimal such theory called "**Q**" (or, sometimes, "Robinson's Q," after Raphael Robinson). We will say what it means for a function to be *representable* in **Q**, and then we will prove the following:

> A function is representable in **Q** if and only if it is computable.

For one thing, this provides us with another model of computability. But we will also use it to show that the set $\{A : \mathbf{Q} \vdash A\}$ is not decidable, by reducing the halting problem to it. By the time we are done, we will have proved much stronger things than this.

The language of **Q** is the language of arithmetic; **Q** consists of the following axioms (to be used in conjunction with the other axioms and rules of first-order logic with identity predicate):

$$\forall x\, \forall y\, (x' = y' \rightarrow x = y) \qquad (Q_1)$$

$$\forall x\, 0 \neq x' \qquad (Q_2)$$

$$\forall x\,(x = \mathsf{o} \lor \exists y\, x = y') \tag{Q_3}$$

$$\forall x\,(x + \mathsf{o}) = x \tag{Q_4}$$

$$\forall x\, \forall y\,(x + y') = (x + y)' \tag{Q_5}$$

$$\forall x\,(x \times \mathsf{o}) = \mathsf{o} \tag{Q_6}$$

$$\forall x\, \forall y\,(x \times y') = ((x \times y) + x) \tag{Q_7}$$

$$\forall x\, \forall y\,(x < y \leftrightarrow \exists z\,(z' + x) = y) \tag{Q_8}$$

For each natural number n, define the numeral $\overline{n}$ to be the term $0''^{\cdots\prime}$ where there are n tick marks in all. So, $\overline{0}$ is the constant symbol o by itself, $\overline{1}$ is o', $\overline{2}$ is o'', etc.

As a theory of arithmetic, **Q** is *extremely* weak; for example, you can't even prove very simple facts like $\forall x\, x \neq x'$ or $\forall x\, \forall y\,(x + y) = (y + x)$. But we will see that much of the reason that **Q** is so interesting is *because* it is so weak. In fact, it is just barely strong enough for the incompleteness theorem to hold. Another reason **Q** is interesting is because it has a *finite* set of axioms.

A stronger theory than **Q** (called *Peano arithmetic* **PA**) is obtained by adding a schema of induction to **Q**:

$$(A(\mathsf{o}) \land \forall x\,(A(x) \rightarrow A(x'))) \rightarrow \forall x\, A(x)$$

where $A(x)$ is any formula. If $A(x)$ contains free variables other than x, we add universal quantifiers to the front to bind all of them (so that the corresponding instance of the induction schema is a sentence). For instance, if $A(x, y)$ also contains the variable y free, the corresponding instance is

$$\forall y\,((A(\mathsf{o}) \land \forall x\,(A(x) \rightarrow A(x'))) \rightarrow \forall x\, A(x))$$

Using instances of the induction schema, one can prove much more from the axioms of **PA** than from those of **Q**. In fact, it takes a good deal of work to find "natural" statements about the natural numbers that can't be proved in Peano arithmetic!

Definition 4.1. A function $f(x_0, \dots, x_k)$ from the natural numbers to the natural numbers is said to be *representable in* **Q** if there is a formula $A_f(x_0, \dots, x_k, y)$ such that whenever $f(n_0, \dots, n_k) = m$, **Q** proves

1. $A_f(\overline{n_0}, \dots, \overline{n_k}, \overline{m})$
2. $\forall y\,(A_f(\overline{n_0}, \dots, \overline{n_k}, y) \to \overline{m} = y)$.

There are other ways of stating the definition; for example, we could equivalently require that **Q** proves $\forall y\,(A_f(\overline{n_0}, \dots, \overline{n_k}, y) \leftrightarrow y = \overline{m})$.

Theorem 4.2. *A function is representable in* **Q** *if and only if it is computable.*

There are two directions to proving the theorem. The left-to-right direction is fairly straightforward once arithmetization of syntax is in place. The other direction requires more work. Here is the basic idea: we pick "general recursive" as a way of making "computable" precise, and show that every general recursive function is representable in **Q**. Recall that a function is general recursive if it can be defined from zero, the successor function succ, and the projection functions P_i^n, using composition, primitive recursion, and regular minimization. So one way of showing that every general recursive function is representable in **Q** is to show that the basic functions are representable, and whenever some functions are representable, then so are the functions defined from them using composition, primitive recursion, and regular minimization. In other words, we might show that the basic functions are representable, and that the representable functions are "closed under" composition, primitive recursion, and regular minimization. This guarantees that every general recursive function is representable.

It turns out that the step where we would show that representable functions are closed under primitive recursion is hard.

In order to avoid this step, we show first that in fact we can do without primitive recursion. That is, we show that every general recursive function can be defined from basic functions using composition and regular minimization alone. To do this, we show that primitive recursion can actually be done by a specific regular minimization. However, for this to work, we have to add some additional basic functions: addition, multiplication, and the characteristic function of the identity relation $\chi_=$. Then, we can prove the theorem by showing that all of *these* basic functions are representable in **Q**, and the representable functions are closed under composition and regular minimization.

4.2 Functions Representable in Q are Computable

Lemma 4.3. *Every function that is representable in* **Q** *is computable.*

Proof. Let's first give the intuitive idea for why this is true. If $f(x_0, \dots, x_k)$ is representable in **Q**, there is a formula $A(x_0, \dots, x_k, y)$ such that

$$\mathbf{Q} \vdash A_f(\overline{n_0}, \dots, \overline{n_k}, \overline{m}) \quad \text{iff} \quad m = f(n_0, \dots, n_k).$$

To compute f, we do the following. List all the possible derivations δ in the language of arithmetic. This is possible to do mechanically. For each one, check if it is a derivation of a formula of the form $A_f(\overline{n_0}, \dots, \overline{n_k}, \overline{m})$. If it is, m must be $= f(n_0, \dots, n_k)$ and we've found the value of f. The search terminates because $\mathbf{Q} \vdash A_f(\overline{n_0}, \dots, \overline{n_k}, \overline{f(n_0, \dots, n_k)})$, so eventually we find a δ of the right sort.

This is not quite precise because our procedure operates on derivations and formulas instead of just on numbers, and we haven't explained exactly why "listing all possible derivations" is mechanically possible. But as we've seen, it is possible to code terms, formulas, and derivations by Gödel numbers. We've also introduced a precise model of computation, the general recursive functions. And we've seen that the relation $\text{Prf}_Q(d, y)$, which holds iff d is the Gödel number of a derivation of the formula with Gödel number x from the axioms of Q, is (primitive) recursive. Other primitive recursive functions we'll need are num (Proposition 3.6) and Subst (Proposition 3.11). From these, it is possible to define f by minimization; thus, f is recursive.

First, define

$$A(n_0, \dots, n_k, m) = \text{Subst}(\text{Subst}(\dots\text{Subst}({}^{\#}A_f{}^{\#}, \text{num}(n_0), {}^{\#}x_0{}^{\#}), \dots), \text{num}(n_k), {}^{\#}x_k{}^{\#}), \text{num}(m), {}^{\#}y^{\#})$$

This looks complicated, but it's just the function $A(n_0, \dots, n_k, m) = {}^{\#}A_f(\overline{n_0}, \dots, \overline{n_k}, \overline{m})^{\#}$.

Now, consider the relation $R(n_0, \dots, n_k, s)$ which holds if $(s)_0$ is the Gödel number of a derivation from Q of $A_f(\overline{n_0}, \dots, \overline{n_k}, \overline{(s)_1})$:

$$R(n_0, \dots, n_k, s) \quad \text{iff} \quad \text{Prf}_Q((s)_0, A(n_0, \dots, n_k, (s)_1)$$

If we can find an s such that $R(n_0, \dots, n_k, s)$ hold, we have found a pair of numbers—$(s)_0$ and (s_1)—such that $(s)_0$ is the Gödel number of a derivation of $A_f(\overline{n_0}, \dots, \overline{n_k}, (s)_1)$. So looking for s is like looking for the pair d and m in the informal proof. And a computable function that "looks for" such an s can be defined by regular minimization. Note that R is regular: for every $n_0, \dots, n_k$, there is a derivation δ of $Q \vdash A_f(\overline{n_0}, \dots, \overline{n_k}, \overline{f(n_0, \dots, n_k)})$, so $R(n_0, \dots, n_k, s)$ holds for $s = \langle {}^{\#}\delta^{\#}, f(n_0, \dots, n_k)\rangle$. So, we can write f as

$$f(n_0, \dots, n_k) = (\mu s\ R(n_0, \dots, n_k, s))_1. \qquad \square$$

4.3 The Beta Function Lemma

In order to show that we can carry out primitive recursion if addition, multiplication, and $\chi_=$ are available, we need to develop functions that handle sequences. (If we had exponentiation as well, our task would be easier.) When we had primitive recursion, we could define things like the "n-th prime," and pick a fairly straightforward coding. But here we do not have primitive recursion—in fact we want to show that we can do primitive recursion using minimization—so we need to be more clever.

Lemma 4.4. *There is a function $\beta(d, i)$ such that for every sequence $a_0, \ldots, a_n$ there is a number d, such that for every $i \leq n$, $\beta(d, i) = a_i$. Moreover, β can be defined from the basic functions using just composition and regular minimization.*

Think of d as coding the sequence $\langle a_0, \ldots, a_n \rangle$, and $\beta(d, i)$ returning the i-th element. (Note that this "coding" does *not* use the prower-of-primes coding we're already familiar with!). The lemma is fairly minimal; it doesn't say we can concatenate sequences or append elements, or even that we can *compute* d from $a_0, \ldots, a_n$ using functions definable by composition and regular minimization. All it says is that there is a "decoding" function such that every sequence is "coded."

The use of the notation β is Gödel's. To repeat, the hard part of proving the lemma is defining a suitable β using the seemingly restricted resources, i.e., using just composition and minimization—however, we're allowed to use addition, multiplication, and $\chi_=$. There are various ways to prove this lemma, but one of the cleanest is still Gödel's original method, which used a number-theoretic fact called the Chinese Remainder theorem.

Definition 4.5. Two natural numbers a and b are *relatively prime* if their greatest common divisor is 1; in other words, they have no other divisors in common.

Definition 4.6. $a \equiv b \mod c$ means $c \mid (a-b)$, i.e., a and b have the same remainder when divided by c.

Here is the *Chinese Remainder theorem:*

Theorem 4.7. *Suppose $x_0, \ldots, x_n$ are (pairwise) relatively prime. Let $y_0, \ldots, y_n$ be any numbers. Then there is a number z such that*

$$\begin{aligned} z &\equiv y_0 \mod x_0 \\ z &\equiv y_1 \mod x_1 \\ &\vdots \\ z &\equiv y_n \mod x_n. \end{aligned}$$

Here is how we will use the Chinese Remainder theorem: if $x_0, \ldots, x_n$ are bigger than $y_0, \ldots, y_n$ respectively, then we can take z to code the sequence $\langle y_0, \ldots, y_n \rangle$. To recover y_i, we need only divide z by x_i and take the remainder. To use this coding, we will need to find suitable values for $x_0, \ldots, x_n$.

A couple of observations will help us in this regard. Given $y_0, \ldots, y_n$, let

$$j = \max(n, y_0, \ldots, y_n) + 1,$$

and let

$$\begin{aligned} x_0 &= 1 + j\,! \\ x_1 &= 1 + 2 \cdot j\,! \\ x_2 &= 1 + 3 \cdot j\,! \\ &\vdots \\ x_n &= 1 + (n+1) \cdot j\,! \end{aligned}$$

Then two things are true:

1. $x_0, \ldots, x_n$ are relatively prime.

2. For each i, $y_i < x_i$.

To see that (1) is true, note that if p is a prime number and $p \mid x_i$ and $p \mid x_k$, then $p \mid 1 + (i+1)j!$ and $p \mid 1 + (k+1)j!$. But then p divides their difference,

$$(1 + (i+1)j!) - (1 + (k+1)j!) = (i-k)j!.$$

Since p divides $1 + (i+1)j!$, it can't divide $j!$ as well (otherwise, the first division would leave a remainder of 1). So p divides $i-k$, since p divides $(i-k)j!$. But $|i-k|$ is at most n, and we have chosen $j > n$, so this implies that $p \mid j!$, again a contradiction. So there is no prime number dividing both x_i and x_k. Clause (2) is easy: we have $y_i < j < j! < x_i$.

Now let us prove the β function lemma. Remember that we can use 0, successor, plus, times, $\chi_=$, projections, and any function defined from them using composition and minimization applied to regular functions. We can also use a relation if its characteristic function is so definable. As before we can show that these relations are closed under boolean combinations and bounded quantification; for example:

1. $\text{not}(x) = \chi_=(x, 0)$
2. $(\min x \leq z)\, R(x,y) = \mu x\ (R(x,y) \vee x = z)$
3. $(\exists x \leq z)\ R(x,y) \Leftrightarrow R((\min x \leq z)\, R(x,y), y)$

We can then show that all of the following are also definable without primitive recursion:

1. The pairing function, $J(x,y) = \frac{1}{2}[(x+y)(x+y+1)] + x$
2. Projections

   $$K(z) = (\min x \leq q)(\exists y \leq z\, [z = J(x,y)])$$

   and

   $$L(z) = (\min y \leq q)(\exists x \leq z\, [z = J(x,y)]).$$

3. $x < y$

4. $x \mid y$

5. The function $\text{rem}(x, y)$ which returns the remainder when y is divided by x

Now define

$$\beta^*(d_0, d_1, i) = \text{rem}(1 + (i+1)d_1, d_0)$$

and

$$\beta(d, i) = \beta^*(K(d), L(d), i).$$

This is the function we need. Given $a_0, \ldots, a_n$, as above, let

$$j = \max(n, a_0, \ldots, a_n) + 1,$$

and let $d_1 = j!$. By the observations above, we know that $1 + d_1, 1 + 2d_1, \ldots, 1 + (n+1)d_1$ are relatively prime and all are bigger than $a_0, \ldots, a_n$. By the Chinese Remainder theorem there is a value d_0 such that for each i,

$$d_0 \equiv a_i \mod (1 + (i+1)d_1)$$

and so (because d_1 is greater than a_i),

$$a_i = \text{rem}(1 + (i+1)d_1, d_0).$$

Let $d = J(d_0, d_1)$. Then for each $i \leq n$, we have

$$\begin{aligned} \beta(d, i) &= \beta^*(d_0, d_1, i) \\ &= \text{rem}(1 + (i+1)d_1, d_0) \\ &= a_i \end{aligned}$$

which is what we need. This completes the proof of the β-function lemma.

4.4 Simulating Primitive Recursion

Now we can show that definition by primitive recursion can be "simulated" by regular minimization using the beta function. Suppose we have $f(\vec{x})$ and $g(\vec{x}, y, z)$. Then the function $h(x, \vec{z})$ defined from f and g by primitive recursion is

$$h(\vec{x}, y) = f(\vec{z})$$
$$h(\vec{x}, y + 1) = g(\vec{x}, y, h(\vec{x}, y)).$$

We need to show that h can be defined from f and g using just composition and regular minimization, using the basic functions and functions defined from them using composition and regular minimization (such as β).

Lemma 4.8. *If h can be defined from f and g using primitive recursion, it can be defined from f, g, the functions* zero, succ, P^n_i, add, mult, $\chi_=$, *using composition and regular minimization.*

Proof. First, define an auxiliary function $\hat{h}(\vec{x}, y)$ which returns the least number d such that d codes a sequence which satisfies

1. $(d)_0 = f(\vec{x})$, and
2. for each $i < x$, $(d)_{i+1} = g(\vec{x}, i, (d)_i)$,

where now $(d)_i$ is short for $\beta(d, i)$. In other words, $\hat{h}$ returns the sequence $\langle h(\vec{x}, 0), h(\vec{x}, 1), \ldots, h(\vec{x}, y)\rangle$. We can write $\hat{h}$ as

$$\hat{h}(\vec{x}, y) = \mu d\ (\beta(d, 0) = f(\vec{x}) \land (\forall i < y)\ \beta(d, i + 1) = g(\vec{x}, i, \beta(d, i)).$$

Note: no primitive recursion is needed here, just minimization. The function we minimize is regular because of the beta function lemma Lemma 4.4.

But now we have

$$h(\vec{x}, y) = \beta(\hat{h}(\vec{x}, y), y),$$

so h can be defined from the basic functions using just composition and regular minimization. □

4.5 Basic Functions are Representable in Q

First we have to show that all the basic functions are representable in **Q**. In the end, we need to show how to assign to each k-ary basic function $f(x_0, \ldots, x_{k-1})$ a formula $A_f(x_0, \ldots, x_{k-1}, y)$ that represents it.

We will be able to represent zero, successor, plus, times, the characteristic function for equality, and projections. In each case, the appropriate representing function is entirely straightforward; for example, zero is represented by the formula $y = \mathrm{o}$, successor is represented by the formula $x_0' = y$, and addition is represented by the formula $(x_0 + x_1) = y$. The work involves showing that **Q** can prove the relevant sentences; for example, saying that addition is represented by the formula above involves showing that for every pair of natural numbers m and n, **Q** proves

$$\overline{n} + \overline{m} = \overline{n+m} \text{ and}$$
$$\forall y\,((\overline{n} + \overline{m}) = y \rightarrow y = \overline{n+m}).$$

Proposition 4.9. *The zero function* $\mathrm{zero}(x) = 0$ *is represented in* **Q** *by* $y = \mathrm{o}$.

Proposition 4.10. *The successor function* $\mathrm{succ}(x) = x + 1$ *is represented in* **Q** *by* $y = x'$.

Proposition 4.11. *The projection function* $P_i^n(x_0, \ldots, x_{n-1}) = x_i$ *is represented in* **Q** *by* $y = x_i$.

Proposition 4.12. *The characteristic function of* $=$,

$$\chi_=(x_0, x_1) = \begin{cases} 1 & \text{if } x_0 = x_1 \\ 0 & \text{otherwise} \end{cases}$$

is represented in $\mathbf{Q}$ *by*

$$(x_0 = x_1 \wedge y = \overline{1}) \vee (x_0 \neq x_1 \wedge y = \overline{0}).$$

The proof requires the following lemma.

Lemma 4.13. *Given natural numbers n and m, if $n \neq m$, then $\mathbf{Q} \vdash \overline{n} \neq \overline{m}$.*

Proof. Use induction on n to show that for every m, if $n \neq m$, then $Q \vdash \overline{n} \neq \overline{m}$.

In the base case, $n = 0$. If m is not equal to 0, then $m = k + 1$ for some natural number k. We have an axiom that says $\forall x\, 0 \neq x'$. By a quantifier axiom, replacing x by $\overline{k}$, we can conclude $0 \neq \overline{k}'$. But $\overline{k}'$ is just $\overline{m}$.

In the induction step, we can assume the claim is true for n, and consider $n + 1$. Let m be any natural number. There are two possibilities: either $m = 0$ or for some k we have $m = k + 1$. The first case is handled as above. In the second case, suppose $n + 1 \neq k + 1$. Then $n \neq k$. By the induction hypothesis for n we have $\mathbf{Q} \vdash \overline{n} \neq \overline{k}$. We have an axiom that says $\forall x\, \forall y\, x' = y' \rightarrow x = y$. Using a quantifier axiom, we have $\overline{n}' = \overline{k}' \rightarrow \overline{n} = \overline{k}$. Using propositional logic, we can conclude, in $\mathbf{Q}$, $\overline{n} \neq \overline{k} \rightarrow \overline{n}' \neq \overline{k}'$. Using modus ponens, we can conclude $\overline{n}' \neq \overline{k}'$, which is what we want, since $\overline{k}'$ is $\overline{m}$. □

Note that the lemma does not say much: in essence it says that $\mathbf{Q}$ can prove that different numerals denote different objects. For example, $\mathbf{Q}$ proves $0'' \neq 0'''$. But showing that this holds in general requires some care. Note also that although we are using induction, it is induction *outside* of $\mathbf{Q}$.

Proof of Proposition 4.12. If $n = m$, then $\overline{n}$ and $\overline{m}$ are the same term, and $\chi_=(n, m) = 1$. But $\mathbf{Q} \vdash (\overline{n} = \overline{m} \wedge \overline{1} = \overline{1})$, so it proves $A_=(\overline{n}, \overline{m}, \overline{1})$. If $n \neq m$, then $\chi_=(n, m) = 0$. By Lemma 4.13, $\mathbf{Q} \vdash \overline{n} \neq \overline{m}$ and so also $(\overline{n} \neq \overline{m} \wedge 0 = 0)$. Thus $\mathbf{Q} \vdash A_=(\overline{n}, \overline{m}, \overline{0})$.

For the second part, we also have two cases. If $n = m$, we have to show that $\mathbf{Q} \vdash \forall(A_=(\overline{n}, \overline{m}, y) \rightarrow y = \overline{1})$. Arguing informally, suppose $A_=(\overline{n}, \overline{m}, y)$, i.e.,

$$(\overline{n} = \overline{n} \land y = \overline{1}) \lor (\overline{n} \neq \overline{n} \land y = \overline{0})$$

The left disjunct implies $y = \overline{1}$ by logic; the right contradicts $\overline{n} = \overline{n}$ which is provable by logic.

Suppose, on the other hand, that $n \neq m$. Then $A_=(\overline{n}, \overline{m}, y)$ is

$$(\overline{n} = \overline{m} \land y = \overline{1}) \lor (\overline{n} \neq \overline{m} \land y = \overline{0})$$

Here, the left disjunct contradicts $\overline{n} \neq \overline{m}$, which is provable in $\mathbf{Q}$ by Lemma 4.13; the right disjunct entails $y = \overline{0}$. □

Proposition 4.14. *The addition function* $\mathrm{add}(x_0, x_1) = x_0 + x_1$ *is represented in* $\mathbf{Q}$ *by*

$$y = (x_0 + x_1).$$

Lemma 4.15. $\mathbf{Q} \vdash (\overline{n} + \overline{m}) = \overline{n + m}$

Proof. We prove this by induction on m. If $m = 0$, the claim is that $\mathbf{Q} \vdash (\overline{n} + 0) = \overline{n}$. This follows by axiom Q_4. Now suppose the claim for m; let's prove the claim for $m + 1$, i.e., prove that $\mathbf{Q} \vdash (\overline{n} + \overline{m + 1}) = \overline{n + m + 1}$. Note that $\overline{m + 1}$ is just $\overline{m}'$, and $\overline{n + m + 1}$ is just $\overline{n + m}'$. By axiom Q_5, $\mathbf{Q} \vdash (\overline{n} + \overline{m}') = (\overline{n} + \overline{m})'$. By induction hypothesis, $\mathbf{Q} \vdash (\overline{n} + \overline{m}) = \overline{n + m}$. So $\mathbf{Q} \vdash (\overline{n} + \overline{m}') = \overline{n + m}'$. □

Proof of Proposition 4.14. The formula $A_{\mathrm{add}}(x_0, x_1, y)$ representing add is $y = (x_0 + x_1)$. First we show that if $\mathrm{add}(n, m) = k$, then $\mathbf{Q} \vdash A_{\mathrm{add}}(\overline{n}, \overline{m}, \overline{k})$, i.e., $\mathbf{Q} \vdash \overline{k} = (\overline{n} + \overline{m})$. But since $k = n + m$, $\overline{k}$ just is $\overline{n + m}$, and we've shown in Lemma 4.15 that $\mathbf{Q} \vdash (\overline{n} + \overline{m}) = \overline{n + m}$.

We also have to show that if $\mathrm{add}(n, m) = k$, then

$$\mathbf{Q} \vdash \forall y\, (A_{\mathrm{add}}(\overline{n}, \overline{m}, y) \rightarrow y = \overline{k}).$$

Suppose we have $(\overline{n} + \overline{m}) = y$. Since

$$\mathbf{Q} \vdash (\overline{n} + \overline{m}) = \overline{n + m},$$

we can replace the left side with $\overline{n + m}$ and get $\overline{n + m} = y$, for arbitrary y. □

Proposition 4.16. *The multiplication function* $\mathrm{mult}(x_0, x_1) = x_0 \cdot x_1$ *is represented in* $\mathbf{Q}$ *by*

$$y = (x_0 \times x_1).$$

Proof. Exercise. □

Lemma 4.17. $\mathbf{Q} \vdash (\overline{n} \times \overline{m}) = \overline{n \cdot m}$

Proof. Exercise. □

Recall that we use $\times$ for the function symbol of the language of arithmetic, and $\cdot$ for the ordinary multiplication operation on numbers. So $\cdot$ can appear between expressions for numbers (such as in $m \cdot n$) while $\times$ appears only between terms of the language of arithmetic (such as in $(\overline{m} \times \overline{n})$). Even more confusingly, $+$ is used for both the function symbol and the addition operation. When it appears between terms—e.g., in $(\overline{n} + \overline{m})$—it is the 2-place function symbol of the language of arithmetic, and when it appears between numbers—e.g., in $n + m$—it is the addition operation. This includes the case $\overline{n + m}$: this is the standard numeral corresponding to the number $n + m$.

4.6 Composition is Representable in $\mathbf{Q}$

Suppose h is defined by

$$h(x_0, \dots, x_{l-1}) = f(g_0(x_0, \dots, x_{l-1}), \dots, g_{k-1}(x_0, \dots, x_{l-1})).$$

where we have already found formulas $A_f, A_{g_0}, \dots, A_{g_{k-1}}$ representing the functions f, and g_0, …, g_{k-1}, respectively. We have to find a formula A_h representing h.

Let's start with a simple case, where all functions are 1-place, i.e., consider $h(x) = f(g(x))$. If $A_f(y, z)$ represents f, and $A_g(x, y)$ represents g, we need a formula $A_h(x, z)$ that represents h. Note that $h(x) = z$ iff there is a y such that both $z = f(y)$ and $y = g(x)$. (If $h(x) = z$, then $g(x)$ is such a y; if such a y exists, then since $y = g(x)$ and $z = f(y)$, $z = f(g(x))$.) This suggests that $\exists y\,(A_g(x, y) \land A_f(y, z))$ is a good candidate for $A_h(x, z)$. We just have to verify that **Q** proves the relevant formulas.

Proposition 4.18. *If $h(n) = m$, then* **Q** $\vdash A_h(\overline{n}, \overline{m})$.

Proof. Suppose $h(n) = m$, i.e., $f(g(n)) = m$. Let $k = g(n)$. Then

$$\mathbf{Q} \vdash A_g(\overline{n}, \overline{k})$$

since A_g represents g, and

$$\mathbf{Q} \vdash A_f(\overline{k}, \overline{m})$$

since A_f represents f. Thus,

$$\mathbf{Q} \vdash A_g(\overline{n}, \overline{k}) \land A_f(\overline{k}, \overline{m})$$

and consequently also

$$\mathbf{Q} \vdash \exists y\,(A_g(\overline{n}, y) \land A_f(y, \overline{m})),$$

i.e., **Q** $\vdash A_h(\overline{n}, \overline{m})$. □

Proposition 4.19. *If $h(n) = m$, then* $\mathbf{Q} \vdash \forall z\,(A_h(\overline{n}, z) \rightarrow z = \overline{m})$.

Proof. Suppose $h(n) = m$, i.e., $f(g(n)) = m$. Let $k = g(n)$. Then

$$\mathbf{Q} \vdash \forall y\,(A_g(\overline{n}, y) \rightarrow y = \overline{k})$$

since A_g represents g, and

$$\mathbf{Q} \vdash \forall z\,(A_f(\overline{k}, z) \rightarrow z = \overline{m})$$

since A_f represents f. Using just a little bit of logic, we can show that also

$$\mathbf{Q} \vdash \forall z\,(\exists y\,(A_g(\overline{n}, y) \wedge A_f(y, z)) \rightarrow z = \overline{m}).$$

i.e., $\mathbf{Q} \vdash \forall y\,(A_h(\overline{n}, y) \rightarrow y = \overline{m})$. □

The same idea works in the more complex case where f and g_i have arity greater than 1.

Proposition 4.20. *If $A_f(y_0, \dots, y_{k-1}, z)$ represents $f(y_0, \dots, y_{k-1})$ in* **Q**, *and $A_{g_i}(x_0, \dots, x_{l-1}, y)$ represents $g_i(x_0, \dots, x_{l-1})$ in* **Q**, *then*

$$\exists y_0, \dots \exists y_{k-1}\,(A_{g_0}(x_0, \dots, x_{l-1}, y_0) \wedge \dots \wedge \\ A_{g_{k-1}}(x_0, \dots, x_{l-1}, y_{k-1}) \wedge A_f(y_0, \dots, y_{k-1}, z))$$

represents

$$h(x_0, \dots, x_{k-1}) = f(g_0(x_0, \dots, x_{k-1}), \dots, g_0(x_0, \dots, x_{k-1})).$$

Proof. Exercise. □

4.7 Regular Minimization is Representable in Q

Let's consider unbounded search. Suppose $g(x, z)$ is regular and representable in **Q**, say by the formula $A_g(x, z, y)$. Let f be defined by $f(z) = \mu x\,[g(x, z) = 0]$. We would like to find a formula $A_f(z, y)$ representing f. The value of $f(z)$ is that number x

which (a) satisfies $g(x, z) = 0$ and (b) is the least such, i.e., for any $w < x$, $g(w, z) \neq 0$. So the following is a natural choice:

$$A_f(z, y) \equiv A_g(y, z, 0) \land \forall w\,(w < y \rightarrow \neg A_g(w, z, 0)).$$

In the general case, of course, we would have to replace z with $z_0, \ldots, z_k$.

The proof, again, will involve some lemmas about things **Q** is strong enough to prove.

Lemma 4.21. *For every constant symbol a and every natural number n,*

$$\mathbf{Q} \vdash (a' + \overline{n}) = (a + \overline{n})'.$$

Proof. The proof is, as usual, by induction on n. In the base case, $n = 0$, we need to show that **Q** proves $(a' + 0) = (a + 0)'$. But we have:

$$\mathbf{Q} \vdash (a' + 0) = a' \quad \text{by axiom } Q_4 \tag{4.1}$$
$$\mathbf{Q} \vdash (a + 0) = a \quad \text{by axiom } Q_4 \tag{4.2}$$
$$\mathbf{Q} \vdash (a + 0)' = a' \quad \text{by eq. (4.2)} \tag{4.3}$$
$$\mathbf{Q} \vdash (a' + 0) = (a + 0)' \quad \text{by eq. (4.1) and eq. (4.3)}$$

In the induction step, we can assume that we have shown that $\mathbf{Q} \vdash (a' + \overline{n}) = (a + \overline{n})'$. Since $\overline{n+1}$ is $\overline{n}'$, we need to show that **Q** proves $(a' + \overline{n}') = (a + \overline{n}')'$. We have:

$$\mathbf{Q} \vdash (a' + \overline{n}') = (a' + \overline{n})' \quad \text{by axiom } Q_5 \tag{4.4}$$
$$\mathbf{Q} \vdash (a' + \overline{n}') = (a + \overline{n}')' \quad \text{inductive hypothesis} \tag{4.5}$$
$$\mathbf{Q} \vdash (a' + \overline{n})' = (a + \overline{n}')' \quad \text{by eq. (4.4) and eq. (4.5).}$$

□

It is again worth mentioning that this is weaker than saying that **Q** proves $\forall x\,\forall y\,(x' + y) = (x + y)'$. Although this sentence is true in N, **Q** does not prove it.

Lemma 4.22. $\mathbf{Q} \vdash \forall x \, \neg x < 0$.

Proof. We give the proof informally (i.e., only giving hints as to how to construct the formal derivation).

We have to prove $\neg a < 0$ for an arbitrary a. By the definition of $<$, we need to prove $\neg \exists y \, (y' + a) = 0$ in $\mathbf{Q}$. We'll assume $\exists y \, (y' + a) = 0$ and prove a contradiction. Suppose $(b' + a) = 0$. Using Q_3, we have that $a = 0 \lor \exists y \, a = y'$. We distinguish cases.

Case 1: $a = 0$ holds. From $(b' + a) = 0$, we have $(b' + 0) = 0$. By axiom Q_4 of $\mathbf{Q}$, we have $(b' + 0) = b'$, and hence $b' = 0$. But by axiom Q_2 we also have $b' \neq 0$, a contradiction.

Case 2: For some c, $a = c'$. But then we have $(b' + c') = 0$. By axiom Q_5, we have $(b' + c)' = 0$, again contradicting axiom Q_2. □

Lemma 4.23. *For every natural number n,*

$$\mathbf{Q} \vdash \forall x \, (x < \overline{n+1} \rightarrow (x = 0 \lor \cdots \lor x = \overline{n})).$$

Proof. We use induction on n. Let us consider the base case, when $n = 0$. In that case, we need to show $a < \overline{1} \rightarrow a = 0$, for arbitrary a. Suppose $a < \overline{1}$. Then by the defining axiom for $<$, we have $\exists y \, (y' + a) = 0'$ (since $\overline{1} \equiv 0'$).

Suppose b has that property, i.e., we have $(b' + a) = 0'$. We need to show $a = 0$. By axiom Q_3, we have either $a = 0$ or that there is a c such that $a = c'$. In the former case, there is nothing to show. So suppose $a = c'$. Then we have $(b' + c') = 0'$. By axiom Q_5 of $\mathbf{Q}$, we have $(b' + c)' = 0'$. By axiom Q_1, we have $(b' + c) = 0$. But this means, by axiom Q_8, that $c < 0$, contradicting Lemma 4.22.

Now for the inductive step. We prove the case for $n + 1$, assuming the case for n. So suppose $a < \overline{n+2}$. Again using Q_3 we can distinguish two cases: $a = 0$ and for some b, $a = c'$. In the first case, $a = 0 \lor \cdots \lor a = \overline{n+1}$ follows trivially. In the second case, we have $c' < \overline{n+2}$, i.e., $c' < \overline{n+1}'$. By axiom Q_8, for some d, $(d' + c') = \overline{n+1}'$. By axiom Q_5, $(d' + c)' = \overline{n+1}'$.

By axiom Q_1, $(d' + c) = \overline{n+1}$, and so $c < \overline{n+1}$ by axiom Q_8. By inductive hypothesis, $c = 0 \lor \cdots \lor c = \overline{n}$. From this, we get $c' = 0' \lor \cdots \lor c' = \overline{n}'$ by logic, and so $a = \overline{1} \lor \cdots \lor a = \overline{n+1}$ since $a = c'$. □

Lemma 4.24. *For every $m \in \mathbb{N}$,*

$$\mathbf{Q} \vdash \forall y\,((y < \overline{m} \lor \overline{m} < y) \lor y = \overline{m}).$$

Proof. By induction on m. First, consider the case $m = 0$. $\mathbf{Q} \vdash \forall y\,(y = 0 \lor \exists z\, y = z')$ by Q_3. Let a be arbitrary. Then either $a = 0$ or for some b, $a = b'$. In the former case, we also have $(a < 0 \lor 0 < a) \lor a = 0$. But if $a = b'$, then $(b' + 0) = (a + 0)$ by the logic of $=$. By Q_4, $(a + 0) = a$, so we have $(b' + 0) = a$, and hence $\exists z\,(z' + 0) = a$. By the definition of $<$ in Q_8, $0 < a$. If $0 < a$, then also $(0 < a \lor a < 0) \lor a = 0$.

Now suppose we have

$$\mathbf{Q} \vdash \forall y\,((y < \overline{m} \lor \overline{m} < y) \lor y = \overline{m})$$

and we want to show

$$\mathbf{Q} \vdash \forall y\,((y < \overline{m+1} \lor \overline{m+1} < y) \lor y = \overline{m+1})$$

Let a be arbitrary. By Q_3, either $a = 0$ or for some b, $a = b'$. In the first case, we have $\overline{m}' + a = \overline{m+1}$ by Q_4, and so $a < \overline{m+1}$ by Q_8.

Now consider the second case, $a = b'$. By the induction hypothesis, $(b < \overline{m} \lor \overline{m} < b) \lor b = \overline{m}$.

The first disjunct $b < \overline{m}$ is equivalent (by Q_8) to $\exists z\,(z'+b) = \overline{m}$. Suppose c has this property. If $(c' + b) = \overline{m}$, then also $(c' + b)' = \overline{m}'$. By Q_5, $(c' + b)' = (c' + b')$. Hence, $(c' + b') = \overline{m}'$. We get $\exists u\,(u'+b') = \overline{m+1}$ by existentially generalizing on c' and keeping in mind that $\overline{m}' \equiv \overline{m+1}$. Hence, if $b < \overline{m}$ then $b' < \overline{m+1}$ and so $a < \overline{m+1}$.

Now suppose $\overline{m} < b$, i.e., $\exists z\,(z' + \overline{m}) = b$. Suppose c is such a z, i.e., $(c' + \overline{m}) = b$. By logic, $(c' + \overline{m})' = b'$. By Q_5, $(c' + \overline{m}') = b'$. Since $a = b'$ and $\overline{m}' \equiv \overline{m+1}$, $(c' + \overline{m+1}) = a$. By Q_8, $\overline{m+1} < a$.

Finally, assume $b = \overline{m}$. Then, by logic, $b' = \overline{m}'$, and so $a = \overline{m+1}$.

Hence, from each disjunct of the case for m and b, we can obtain the corresponding disjunct for for $m + 1$ and a. □

Proposition 4.25. *If $A_g(x, z, y)$ represents $g(x, y)$ in $\mathbf{Q}$, then*

$$A_f(z, y) \equiv A_g(y, z, 0) \land \forall w\,(w < y \rightarrow \neg A_g(w, z, 0)).$$

represents $f(z) = \mu x\,[g(x, z) = 0]$.

Proof. First we show that if $f(n) = m$, then $\mathbf{Q} \vdash A_f(\overline{n}, \overline{m})$, i.e.,

$$\mathbf{Q} \vdash A_g(\overline{m}, \overline{n}, 0) \land \forall w\,(w < \overline{m} \rightarrow \neg A_g(w, \overline{n}, 0)).$$

Since $A_g(x, z, y)$ represents $g(x, z)$ and $g(m, n) = 0$ if $f(n) = m$, we have

$$\mathbf{Q} \vdash A_g(\overline{m}, \overline{n}, 0).$$

If $f(n) = m$, then for every $k < m$, $g(k, n) \neq 0$. So

$$\mathbf{Q} \vdash \neg A_g(\overline{k}, \overline{n}, 0).$$

We get that

$$\mathbf{Q} \vdash \forall w\,(w < \overline{m} \rightarrow \neg A_g(w, \overline{n}, 0)). \tag{4.6}$$

by Lemma 4.22 in case $m = 0$ and by Lemma 4.23 otherwise.

Now let's show that if $f(n) = m$, then $\mathbf{Q} \vdash \forall y\,(A_f(\overline{n}, y) \rightarrow y = \overline{m})$. We again sketch the argument informally, leaving the formalization to the reader.

Suppose $A_f(\overline{n}, b)$. From this we get (a) $A_g(b, \overline{n}, 0)$ and (b) $\forall w\,(w < b \rightarrow \neg A_g(w, \overline{n}, 0))$. By Lemma 4.24, $(b < \overline{m} \lor \overline{m} < b) \lor b = \overline{m}$. We'll show that both $b < \overline{m}$ and $\overline{m} < b$ leads to a contradiction.

If $\overline{m} < b$, then $\neg A_g(\overline{m}, \overline{n}, 0)$ from (b). But $m = f(n)$, so $g(m, n) = 0$, and so $\mathbf{Q} \vdash A_g(\overline{m}, \overline{n}, 0)$ since A_g represents g. So we have a contradiction.

Now suppose $b < \overline{m}$. Then since $\mathbf{Q} \vdash \forall w\,(w < \overline{m} \rightarrow \neg A_g(w, \overline{n}, 0))$ by eq. (4.6), we get $\neg A_g(b, \overline{n}, 0)$. This again contradicts (a). □

4.8 Computable Functions are Representable in Q

Theorem 4.26. *Every computable function is representable in* **Q**.

Proof. For definiteness, and using the Church-Turing Thesis, let's say that a function is computable iff it is general recursive. The general recursive functions are those which can be defined from the zero function zero, the successor function succ, and the projection function P_i^n using composition, primitive recursion, and regular minimization. By Lemma 4.8, any function h that can be defined from f and g can also be defined using composition and regular minimization from f, g, and zero, succ, P_i^n, add, mult, $\chi_=$. Consequently, a function is general recursive iff it can be defined from zero, succ, P_i^n, add, mult, $\chi_=$ using composition and regular minimization.

We've furthermore shown that the basic functions in question are representable in **Q** (Propositions 4.9 to 4.12, 4.14 and 4.16), and that any function defined from representable functions by composition or regular minimization (Proposition 4.20, Proposition 4.25) is also representable. Thus every general recursive function is representable in **Q**. □

We have shown that the set of computable functions can be characterized as the set of functions representable in **Q**. In fact, the proof is more general. From the definition of representability, it is not hard to see that any theory extending **Q** (or in which one can interpret **Q**) can represent the computable functions. But, conversely, in any proof system in which the notion of proof is computable, every representable function is computable. So, for example, the set of computable functions can be characterized as the set of functions representable in Peano arithmetic, or even Zermelo-Fraenkel set theory. As Gödel noted, this is somewhat surprising. We will see that when it comes to provability, questions are very sensitive to which theory you consider; roughly, the stronger the axioms, the more you can prove. But across a wide range of axiomatic theories, the representable functions are exactly the computable ones; stronger theories do not represent more functions as long as they are axiomatizable.

4.9 Representing Relations

Let us say what it means for a *relation* to be representable.

Definition 4.27. A relation $R(x_0, \ldots, x_k)$ on the natural numbers is *representable in* **Q** if there is a formula $A_R(x_0, \ldots, x_k)$ such that whenever $R(n_0, \ldots, n_k)$ is true, **Q** proves $A_R(\overline{n_0}, \ldots, \overline{n_k})$, and whenever $R(n_0, \ldots, n_k)$ is false, **Q** proves $\neg A_R(\overline{n_0}, \ldots, \overline{n_k})$.

Theorem 4.28. *A relation is representable in* **Q** *if and only if it is computable.*

Proof. For the forwards direction, suppose $R(x_0, \ldots, x_k)$ is represented by the formula $A_R(x_0, \ldots, x_k)$. Here is an algorithm for computing R: on input $n_0, \ldots, n_k$, simultaneously search for a proof of $A_R(\overline{n_0}, \ldots, \overline{n_k})$ and a proof of $\neg A_R(\overline{n_0}, \ldots, \overline{n_k})$. By our hypothesis, the search is bound to find one or the other; if it is the first, report "yes," and otherwise, report "no."

In the other direction, suppose $R(x_0, \ldots, x_k)$ is computable. By definition, this means that the function $\chi_R(x_0, \ldots, x_k)$ is computable. By Theorem 4.2, χ_R is represented by a formula, say $A_{\chi_R}(x_0, \dots, x_k, y)$. Let $A_R(x_0, \dots, x_k)$ be the formula $A_{\chi_R}(x_0, \dots, x_k, \overline{1})$. Then for any n_0, $\dots$, n_k, if $R(n_0, \dots, n_k)$ is true, then $\chi_R(n_0, \dots, n_k) = 1$, in which case **Q** proves $A_{\chi_R}(\overline{n_0}, \dots, \overline{n_k}, \overline{1})$, and so **Q** proves $A_R(\overline{n_0}, \dots, \overline{n_k})$. On the other hand, if $R(n_0, \dots, n_k)$ is false, then $\chi_R(n_0, \dots, n_k) = 0$. This means that **Q** proves

$$\forall y\, (A_{\chi_R}(\overline{n_0}, \dots, \overline{n_k}, y) \rightarrow y = \overline{0}).$$

Since **Q** proves $\overline{0} \neq \overline{1}$, **Q** proves $\neg A_{\chi_R}(\overline{n_0}, \dots, \overline{n_k}, \overline{1})$, and so it proves $\neg A_R(\overline{n_0}, \dots, \overline{n_k})$. □

4.10 Undecidability

We call a theory **T** *undecidable* if there is no computational procedure which, after finitely many steps and unfailingly, provides a correct answer to the question "does **T** prove A?" for any sentence A in the language of **T**. So **Q** would be decidable iff there were a computational procedure which decides, given a sentence A in the language of arithmetic, whether $\mathbf{Q} \vdash A$ or not. We can make this more precise by asking: Is the relation $\mathrm{Prov}_{\mathbf{Q}}(y)$, which holds of y iff y is the Gödel number of a sentence provable in **Q**, recursive? The answer is: no.

Theorem 4.29. *Q is undecidable, i.e., the relation*

$$\mathrm{Prov}_{\mathbf{Q}}(y) \Leftrightarrow \mathrm{Sent}(y) \wedge \exists x\, \mathrm{Prf}_{\mathbf{Q}}(x, y)$$

is not recursive.

Proof. Suppose it were. Then we could solve the halting problem as follows: Given e and n, we know that $\varphi_e(n) \downarrow$ iff there is an s such that $T(e, n, s)$, where T is Kleene's predicate from Theorem 2.28. Since T is primitive recursive it is representable in **Q** by a formula B_T, that is, $\mathbf{Q} \vdash B_T(\overline{e}, \overline{n}, \overline{s})$ iff $T(e, n, s)$. If $\mathbf{Q} \vdash B_T(\overline{e}, \overline{n}, \overline{s})$ then also $\mathbf{Q} \vdash \exists y\, B_T(\overline{e}, \overline{n}, y)$. If no such s exists, then $\mathbf{Q} \vdash \neg B_T(\overline{e}, \overline{n}, \overline{s})$ for every s. But $\mathbf{Q}$ is ω-consistent, i.e., if $\mathbf{Q} \vdash \neg A(\overline{n})$ for every $n \in \mathbb{N}$, then $\mathbf{Q} \nvdash \exists y\, A(y)$. We know this because the axioms of $\mathbf{Q}$ are true in the standard model N. So, $\mathbf{Q} \nvdash \exists y\, B_T(\overline{e}, \overline{n}, y)$. In other words, $\mathbf{Q} \vdash \exists y\, B_T(\overline{e}, \overline{n}, y)$ iff there is an s such that $T(e, n, s)$, i.e., iff $\varphi_e(n) \downarrow$. From e and n we can compute ${}^{\#}\exists y\, B_T(\overline{e}, \overline{n}, y)^{\#}$, let $g(e, n)$ be the primitive recursive function which does that. So

$$h(e, n) = \begin{cases} 1 & \text{if } \mathrm{Prov}_{\mathbf{Q}}(g(e, n)) \\ 0 & \text{otherwise.} \end{cases}$$

This would show that h is recursive if $\mathrm{Prov}_{\mathbf{Q}}$ is. But h is not recursive, by Theorem 2.29, so $\mathrm{Prov}_{\mathbf{Q}}$ cannot be either. □

Corollary 4.30. *First-order logic is undecidable.*

Proof. If first-order logic were decidable, provability in $\mathbf{Q}$ would be as well, since $\mathbf{Q} \vdash A$ iff $\vdash O \to A$, where O is the conjunction of the axioms of $\mathbf{Q}$. □

Summary

In order to show how theories like $\mathbf{Q}$ can "talk" about computable functions—and especially about provability (via Gödel numbers)—we established that $\mathbf{Q}$ **represents** all computable functions. By "$\mathbf{Q}$ represents $f(n)$" we mean that there is a formula $A_f(x, y)$ in $\mathcal{L}_A$ which expresses that $f(x) = y$, and $\mathbf{Q}$ can prove that it does. This, in turn, means that whenever $f(n) = m$, then $\mathbf{T} \vdash A_f(\overline{n}, \overline{m})$ and $\mathbf{T} \vdash \forall y\, (A_f(\overline{n}, y) \to y = \overline{m})$. (Here, $\overline{n}$ is the **standard numeral** for n, i.e., the term $0'^{\cdots\prime}$ with n $\prime$s. The term $\overline{n}$ picks out the number n in the standard model N, so it's a convenient way of representing the number n in $\mathcal{L}_A$.) To prove that $\mathbf{Q}$ represents all computable functions we go back to the characterization of computable functions as those that can be defined from zero, succ, and the projection functions, by composition, primitive recursion, and regular minimization. While it is relatively easy to prove that the basic functions are representable and that functions defined by composition and regular minimization from representable functions are also representable, primitive recursion is harder. We showed that we can actually avoid definition by primitive recursion, if we allow a few additional basic functions (namely, addition, multiplication, and the characteristic function of =). This required a **beta function** which allows us to deal with sequences of numbers in a rudimentary way, and which can be defined without using primitive recursion.

Problems

Problem 4.1. Prove that $y = 0$, $y = x'$, and $y = x_i$ represent zero, succ, and P_i^n, respectively.

Problem 4.2. Prove Lemma 4.17.

Problem 4.3. Use Lemma 4.17 to prove Proposition 4.16.

Problem 4.4. Using the proofs of Proposition 4.19 and Proposition 4.19 as a guide, carry out the proof of Proposition 4.20 in detail.

Problem 4.5. Show that if R is representable in **Q**, so is χ_R.

CHAPTER 5

Incompleteness and Provability

5.1 Introduction

Hilbert thought that a system of axioms for a mathematical structure, such as the natural numbers, is inadequate unless it allows one to derive all true statements about the structure. Combined with his later interest in formal systems of deduction, this suggests that he thought that we should guarantee that, say, the formal systems we are using to reason about the natural numbers is not only consistent, but also *complete*, i.e., every statement in its language is either derivable or its negation is. Gödel's first incompleteness theorem shows that no such system of axioms exists: there is no complete, consistent, axiomatizable formal system for arithmetic. In fact, no "sufficiently strong," consistent, axiomatizable mathematical theory is complete.

A more important goal of Hilbert's, the centerpiece of his program for the justification of modern ("classical") mathematics, was to find finitary consistency proofs for formal systems representing classical reasoning. With regard to Hilbert's program, then, Gödel's second incompleteness theorem was a much bigger blow. The second incompleteness theorem can be stated in vague terms, like the first incompleteness theorem. Roughly speaking, it says that no sufficiently strong theory of arithmetic can prove its own consistency. We will have to take "sufficiently strong" to include a little bit more than **Q**.

The idea behind Gödel's original proof of the incompleteness theorem can be found in the Epimenides paradox. Epimenides, a Cretan, asserted that all Cretans are liars; a more direct form of the paradox is the assertion "this sentence is false." Essentially, by replacing truth with derivability, Gödel was able to formalize a sentence which, in a roundabout way, asserts that it itself is not derivable. If that sentence were derivable, the theory would then be inconsistent. Gödel showed that the negation of that sentence is also not derivable from the system of axioms he was considering. (For this second part, Gödel had to assume that the theory **T** is what's called "ω-consistent." ω-Consistency is related to consistency, but is a stronger property. A few years after Gödel, Rosser showed that assuming simple consistency of **T** is enough.)

The first challenge is to understand how one can construct a sentence that refers to itself. For every formula A in the language of **Q**, let $\ulcorner A \urcorner$ denote the numeral corresponding to ${}^{\#}A^{\#}$. Think about what this means: A is a formula in the language of **Q**, ${}^{\#}A^{\#}$ is a natural number, and $\ulcorner A \urcorner$ is a *term* in the language of **Q**. So every formula A in the language of **Q** has a *name*, $\ulcorner A \urcorner$, which is a term in the language of **Q**; this provides us with a conceptual framework in which formulas in the language of **Q** can "say" things about other formulas. The following lemma is known as the fixed-point lemma.

Lemma 5.1. *Let* **T** *be any theory extending* **Q**, *and let* $B(x)$ *be any formula with only the variable* x *free. Then there is a sentence* A *such that* $\mathbf{T} \vdash A \leftrightarrow B(\ulcorner A \urcorner)$.

The lemma asserts that given any property $B(x)$, there is a sentence A that asserts "$B(x)$ is true of me," and **T** "knows" this.

How can we construct such a sentence? Consider the following version of the Epimenides paradox, due to Quine:

> "Yields falsehood when preceded by its quotation" yields falsehood when preceded by its quotation.

This sentence is not directly self-referential. It simply makes an assertion about the syntactic objects between quotes, and, in doing so, it is on par with sentences like

1. "Robert" is a nice name.
2. "I ran." is a short sentence.
3. "Has three words" has three words.

But what happens when one takes the phrase "yields falsehood when preceded by its quotation," and precedes it with a quoted version of itself? Then one has the original sentence! In short, the sentence asserts that it is false.

5.2 The Fixed-Point Lemma

The fixed-point lemma says that for any formula $B(x)$, there is a sentence A such that $\mathbf{T} \vdash A \leftrightarrow B(\ulcorner A \urcorner)$, provided $\mathbf{T}$ extends $\mathbf{Q}$. In the case of the liar sentence, we'd want A to be equivalent (provably in $\mathbf{T}$) to "$\ulcorner A \urcorner$ is false," i.e., the statement that $^{\#}A^{\#}$ is the Gödel number of a false sentence. To understand the idea of the proof, it will be useful to compare it with Quine's informal gloss of A as, "'yields a falsehood when preceded by its own quotation' yields a falsehood when preceded by its own quotation." The operation of taking an expression, and then forming a sentence by preceding this expression by its own quotation may be called *diagonalizing* the expression, and the result its diagonalization. So, the diagonalization of 'yields a falsehood when preceded by its own quotation' is "'yields a falsehood when preceded by its own quotation' yields a falsehood when preceded by its own quotation." Now note that Quine's liar sentence is not the diagonalization of 'yields a falsehood' but of 'yields a falsehood when preceded by its own quotation.' So the property being diagonalized to yield the liar sentence itself involves diagonalization!

In the language of arithmetic, we form quotations of a formula with one free variable by computing its Gödel numbers and then substituting the standard numeral for that Gödel number into the free variable. The diagonalization of $E(x)$ is $E(\overline{n})$, where $n = {}^{\#}E(x)^{\#}$. (From now on, let's abbreviate $\overline{{}^{\#}E(x)^{\#}}$ as $\ulcorner E(x) \urcorner$.) So if $B(x)$ is "is a falsehood," then "yields a falsehood if preceded by its own quotation," would be "yields a falsehood when applied to the Gödel number of its diagonalization." If we had a symbol diag for the function $\mathrm{diag}(n)$ which computes the Gödel number of the diagonalization of the formula with Gödel number n, we could write $E(x)$ as $B(\mathit{diag}(x))$. And Quine's version of the liar sentence would then be the diagonalization of it, i.e., $E(\ulcorner E \urcorner)$ or $B(\mathit{diag}(\ulcorner B(\mathit{diag}(x)) \urcorner))$. Of course, $B(x)$ could now be any other property, and the same construction would work. For the incompleteness theorem, we'll take $B(x)$ to be "x is not derivable in $\mathbf{T}$." Then $E(x)$ would be "yields a sentence not derivable in $\mathbf{T}$ when applied to the Gödel number of its diagonalization."

To formalize this in $\mathbf{T}$, we have to find a way to formalize diag. The function $\mathrm{diag}(n)$ is computable, in fact, it is primitive recursive: if n is the Gödel number of a formula $E(x)$, $\mathrm{diag}(n)$ returns the Gödel number of $E(\ulcorner E(x) \urcorner)$. (Recall, $\ulcorner E(x) \urcorner$ is the standard numeral of the Gödel number of $E(x)$, i.e., $\overline{{}^{\#}E(x)^{\#}}$). If diag were a function symbol in $\mathbf{T}$ representing the function diag, we could take A to be the formula $B(\mathit{diag}(\ulcorner B(\mathit{diag}(x)) \urcorner))$. Notice that

$$\begin{aligned} \mathrm{diag}({}^{\#}B(\mathit{diag}(x))^{\#}) &= {}^{\#}B(\mathit{diag}(\ulcorner B(\mathit{diag}(x)) \urcorner)^{\#} \\ &= {}^{\#}A^{\#}. \end{aligned}$$

Assuming $\mathbf{T}$ can derive

$$\mathit{diag}(\ulcorner B(\mathit{diag}(x)) \urcorner) = \ulcorner A \urcorner,$$

it can derive $B(\mathit{diag}(\ulcorner B(\mathit{diag}(x)) \urcorner)) \leftrightarrow B(\ulcorner A \urcorner)$. But the left hand side is, by definition, A.

Of course, *diag* will in general not be a function symbol of **T**, and certainly is not one of **Q**. But, since diag is computable, it is *representable* in **Q** by some formula $D_{\text{diag}}(x, y)$. So instead of writing $B(diag(x))$ we can write $\exists y\,(D_{\text{diag}}(x, y) \land B(y))$. Otherwise, the proof sketched above goes through, and in fact, it goes through already in **Q**.

Lemma 5.2. *Let $B(x)$ be any formula with one free variable x. Then there is a sentence A such that $\mathbf{Q} \vdash A \leftrightarrow B(\ulcorner A \urcorner)$.*

Proof. Given $B(x)$, let $E(x)$ be the formula $\exists y\,(D_{\text{diag}}(x, y) \land B(y))$ and let A be its diagonalization, i.e., the formula $E(\ulcorner E(x) \urcorner)$.

Since D_{diag} represents diag, and $\text{diag}({}^{\#}E(x)^{\#}) = {}^{\#}A^{\#}$, **Q** can derive

$$D_{\text{diag}}(\ulcorner E(x) \urcorner, \ulcorner A \urcorner) \tag{5.1}$$

$$\forall y\,(D_{\text{diag}}(\ulcorner E(x) \urcorner, y) \rightarrow y = \ulcorner A \urcorner). \tag{5.2}$$

Now we show that $\mathbf{Q} \vdash A \leftrightarrow B(\ulcorner A \urcorner)$. We argue informally, using just logic and facts derivable in **Q**.

First, suppose A, i.e., $E(\ulcorner E(x) \urcorner)$. Going back to the definition of $E(x)$, we see that $E(\ulcorner E(x) \urcorner)$ just is

$$\exists y\,(D_{\text{diag}}(\ulcorner E(x) \urcorner, y) \land B(y)).$$

Consider such a y. Since $D_{\text{diag}}(\ulcorner E(x) \urcorner, y)$, by eq. (5.2), $y = \ulcorner A \urcorner$. So, from $B(y)$ we have $B(\ulcorner A \urcorner)$.

Now suppose $B(\ulcorner A \urcorner)$. By eq. (5.1), we have $D_{\text{diag}}(\ulcorner E(x) \urcorner, \ulcorner A \urcorner) \land B(\ulcorner A \urcorner)$. It follows that $\exists y\,(D_{\text{diag}}(\ulcorner E(x) \urcorner, y) \land B(y))$. But that's just $E(\ulcorner E \urcorner)$, i.e., A. □

You should compare this to the proof of the fixed-point lemma in computability theory. The difference is that here we want to define a *statement* in terms of itself, whereas there we wanted to define a *function* in terms of itself; this difference aside, it is really the same idea.

5.3 The First Incompleteness Theorem

We can now describe Gödel's original proof of the first incompleteness theorem. Let **T** be any computably axiomatized theory in a language extending the language of arithmetic, such that **T** includes the axioms of **Q**. This means that, in particular, **T** represents computable functions and relations.

We have argued that, given a reasonable coding of formulas and proofs as numbers, the relation $\mathrm{Prf}_T(x,y)$ is computable, where $\mathrm{Prf}_T(x,y)$ holds if and only if x is the Gödel number of a derivation of the formula with Gödel number y in **T**. In fact, for the particular theory that Gödel had in mind, Gödel was able to show that this relation is primitive recursive, using the list of 45 functions and relations in his paper. The 45th relation, xBy, is just $\mathrm{Prf}_T(x,y)$ for his particular choice of **T**. Remember that where Gödel uses the word "recursive" in his paper, we would now use the phrase "primitive recursive."

Since $\mathrm{Prf}_T(x,y)$ is computable, it is representable in **T**. We will use $\mathsf{Prf}_T(x,y)$ to refer to the formula that represents it. Let $\mathsf{Prov}_T(y)$ be the formula $\exists x\, \mathsf{Prf}_T(x,y)$. This describes the 46th relation, $\mathrm{Bew}(y)$, on Gödel's list. As Gödel notes, this is the only relation that "cannot be asserted to be recursive." What he probably meant is this: from the definition, it is not clear that it is computable; and later developments, in fact, show that it isn't.

Let **T** be an axiomatizable theory containing **Q**. Then $\mathrm{Prf}_T(x,y)$ is decidable, hence representable in **Q** by a formula $\mathsf{Prf}_T(x,y)$. Let $\mathsf{Prov}_T(y)$ be the formula we described above. By the fixed-point lemma, there is a formula $G_{\mathbf{T}}$ such that **Q** (and hence **T**) derives

$$G_{\mathbf{T}} \leftrightarrow \neg\mathsf{Prov}_T(\ulcorner G_{\mathbf{T}} \urcorner). \tag{5.3}$$

Note that $G_{\mathbf{T}}$ says, in essence, "$G_{\mathbf{T}}$ is not derivable in **T**."

Lemma 5.3. *If* **T** *is a consistent, axiomatizable theory extending* **Q**, *then* $\mathbf{T} \nvdash G_{\mathbf{T}}$.

Proof. Suppose **T** derives $G_{\mathbf{T}}$. Then there *is* a derivation, and so, for some number m, the relation $\mathrm{Prf}_T(m, {}^{\#}G_{\mathbf{T}}{}^{\#})$ holds. But then **Q** derives the sentence $\mathsf{Prf}_T(\overline{m}, \ulcorner G_{\mathbf{T}} \urcorner)$. So **Q** derives $\exists x\, \mathsf{Prf}_T(x, \ulcorner G_{\mathbf{T}} \urcorner)$, which is, by definition, $\mathsf{Prov}_T(\ulcorner G_{\mathbf{T}} \urcorner)$. By eq. (5.3), **Q** derives $\neg G_{\mathbf{T}}$, and since **T** extends **Q**, so does **T**. We have shown that if **T** derives $G_{\mathbf{T}}$, then it also derives $\neg G_{\mathbf{T}}$, and hence it would be inconsistent. □

Definition 5.4. A theory **T** is *ω-consistent* if the following holds: if $\exists x\, A(x)$ is any sentence and **T** derives $\neg A(\overline{0})$, $\neg A(\overline{1})$, $\neg A(\overline{2})$, ... then **T** does not prove $\exists x\, A(x)$.

Note that every ω-consistent theory is also consistent. This follows simply from the fact that if **T** is inconsistent, then $\mathbf{T} \vdash A$ for every A. In particular, if **T** is inconsistent, it derives both $\neg A(\overline{n})$ for every n and also derives $\exists x\, A(x)$. So, if **T** is inconsistent, it is ω-inconsistent. By contraposition, if **T** is ω-consistent, it must be consistent.

Lemma 5.5. *If* **T** *is an ω-consistent, axiomatizable theory extending* **Q**, *then* $\mathbf{T} \nvdash G_{\mathbf{T}}$.

Proof. We show that if **T** derives $\neg G_{\mathbf{T}}$, then it is ω-inconsistent. Suppose **T** derives $\neg G_{\mathbf{T}}$. If **T** is inconsistent, it is ω-inconsistent, and we are done. Otherwise, **T** is consistent, so it does not derive $G_{\mathbf{T}}$ by Lemma 5.3. Since there is no derivation of $G_{\mathbf{T}}$ in **T**, **Q** derives

$$\neg\mathsf{Prf}_T(\overline{0}, \ulcorner G_{\mathbf{T}} \urcorner), \neg\mathsf{Prf}_T(\overline{1}, \ulcorner G_{\mathbf{T}} \urcorner), \neg\mathsf{Prf}_T(\overline{2}, \ulcorner G_{\mathbf{T}} \urcorner), \ldots$$

and so does **T**. On the other hand, by eq. (5.3), $\neg G_{\mathbf{T}}$ is equivalent to $\exists x\, \mathsf{Prf}_T(x, \ulcorner G_{\mathbf{T}} \urcorner)$. So **T** is ω-inconsistent. □

Theorem 5.6. *Let* **T** *be any* ω*-consistent, axiomatizable theory extending* **Q**. *Then* **T** *is not complete.*

Proof. If **T** is ω-consistent, it is consistent, so $\mathbf{T} \nvdash G_{\mathbf{T}}$ by Lemma 5.3. By Lemma 5.5, $\mathbf{T} \nvdash \neg G_{\mathbf{T}}$. This means that **T** is incomplete, since it derives neither $G_{\mathbf{T}}$ nor $\neg G_{\mathbf{T}}$. □

5.4 Rosser's Theorem

Can we modify Gödel's proof to get a stronger result, replacing "ω-consistent" with simply "consistent"? The answer is "yes," using a trick discovered by Rosser. Rosser's trick is to use a "modified" derivability predicate $\mathsf{RProv}_T(y)$ instead of $\mathsf{Prov}_T(y)$.

Theorem 5.7. *Let* **T** *be any consistent, axiomatizable theory extending* **Q**. *Then* **T** *is not complete.*

Proof. Recall that $\mathsf{Prov}_T(y)$ is defined as $\exists x\, \mathsf{Prf}_T(x,y)$, where $\mathsf{Prf}_T(x,y)$ represents the decidable relation which holds iff x is the Gödel number of a derivation of the sentence with Gödel number y. The relation that holds between x and y if x is the Gödel number of a *refutation* of the sentence with Gödel number y is also decidable. Let $\mathrm{not}(x)$ be the primitive recursive function which does the following: if x is the code of a formula A, $\mathrm{not}(x)$ is a code of $\neg A$. Then $\mathrm{Ref}_T(x,y)$ holds iff $\mathrm{Prf}_T(x, \mathrm{not}(y))$. Let $\mathsf{Ref}_T(x,y)$ represent it. Then, if $\mathbf{T} \vdash \neg A$ and δ is a corresponding derivation, $\mathbf{Q} \vdash \mathsf{Ref}_T(\ulcorner \delta \urcorner, \ulcorner A \urcorner)$. We define $\mathsf{RProv}_T(y)$ as

$$\exists x\, (\mathsf{Prf}_T(x,y) \land \forall z\, (z < x \rightarrow \neg \mathsf{Ref}_T(z,y))).$$

Roughly, $\mathsf{RProv}_T(y)$ says "there is a proof of y in **T**, and there is no shorter refutation of y." Assuming **T** is consistent, $\mathsf{RProv}_T(y)$ is true of the same numbers as $\mathsf{Prov}_T(y)$; but from the point of view of *provability* in **T** (and we now know that there is a difference between truth and provability!) the two have different properties. If **T** is *in*consistent, then the two do *not* hold of the same numbers! ($\mathsf{RProv}_T(y)$ is often read as "y is Rosser provable." Since, as just discussed, Rosser provability is not some special kind of provability—in inconsistent theories, there are sentences that are provable but not Rosser provable—this may be confusing. To avoid the confusion, you could instead read it as "y is shmovable.")

By the fixed-point lemma, there is a formula $R_{\mathbf{T}}$ such that

$$\mathbf{Q} \vdash R_{\mathbf{T}} \leftrightarrow \neg\mathsf{RProv}_T(\ulcorner R_{\mathbf{T}} \urcorner). \tag{5.4}$$

In contrast to the proof of Theorem 5.6, here we claim that if $\mathbf{T}$ is consistent, $\mathbf{T}$ doesn't derive $R_{\mathbf{T}}$, and $\mathbf{T}$ also doesn't derive $\neg R_{\mathbf{T}}$. (In other words, we don't need the assumption of ω-consistency.)

First, let's show that $\mathbf{T} \nvdash R_T$. Suppose it did, so there is a derivation of R_T from T; let n be its Gödel number. Then $\mathbf{Q} \vdash \mathsf{Prf}_T(\overline{n}, \ulcorner R_T \urcorner)$, since Prf_T represents Prf_T in $\mathbf{Q}$. Also, for each $k < n$, k is not the Gödel number of $\neg R_T$, since $\mathbf{T}$ is consistent. So for each $k < n$, $\mathbf{Q} \vdash \neg\mathsf{Ref}_T(\overline{k}, \ulcorner R_T \urcorner)$. By Lemma 4.23, $\mathbf{Q} \vdash \forall z\,(z < \overline{n} \rightarrow \neg\mathsf{Ref}_T(z, \ulcorner R_T \urcorner))$. Thus,

$$\mathbf{Q} \vdash \exists x\,(\mathsf{Prf}_T(x, \ulcorner R_T \urcorner) \wedge \forall z\,(z < x \rightarrow \neg\mathsf{Ref}_T(z, \ulcorner R_T \urcorner))),$$

but that's just $\mathsf{RProv}_T(\ulcorner R_T \urcorner)$. By eq. (5.4), $\mathbf{Q} \vdash \neg R_T$. Since $\mathbf{T}$ extends $\mathbf{Q}$, also $\mathbf{T} \vdash \neg R_T$. We've assumed that $\mathbf{T} \vdash R_T$, so $\mathbf{T}$ would be inconsistent, contrary to the assumption of the theorem.

Now, let's show that $\mathbf{T} \nvdash \neg R_T$. Again, suppose it did, and suppose n is the Gödel number of a derivation of $\neg R_T$. Then $\mathrm{Ref}_T(n, {}^{\#}R_T{}^{\#})$ holds, and since Ref_T represents Ref_T in $\mathbf{Q}$, $\mathbf{Q} \vdash \mathsf{Ref}_T(\overline{n}, \ulcorner R_T \urcorner)$. We'll again show that $\mathbf{T}$ would then be inconsistent because it would also derive R_T. Since

$$\mathbf{Q} \vdash R_T \leftrightarrow \neg\mathsf{RProv}_T(\ulcorner R_T \urcorner),$$

and since $\mathbf{T}$ extends $\mathbf{Q}$, it suffices to show that

$$\mathbf{Q} \vdash \neg\mathsf{RProv}_T(\ulcorner R_T \urcorner).$$

The sentence $\neg\mathsf{RProv}_T(\ulcorner R_T \urcorner)$, i.e.,

$$\neg\exists x\,(\mathsf{Prf}_T(x, \ulcorner R_T \urcorner) \wedge \forall z\,(z < x \rightarrow \neg\mathsf{Ref}_T(z, \ulcorner R_T \urcorner)))$$

is logically equivalent to

$$\forall x\,(\mathsf{Prf}_T(x, \ulcorner R_T \urcorner) \rightarrow \exists z\,(z < x \land \mathsf{Ref}_T(z, \ulcorner R_T \urcorner)))$$

We argue informally using logic, making use of facts about what **Q** derives. Suppose x is arbitrary and $\mathsf{Prf}_T(x, \ulcorner R_T \urcorner)$. We already know that $\mathbf{T} \nvdash R_T$, and so for every k, $\mathbf{Q} \vdash \lnot\mathsf{Prf}_T(\overline{k}, \ulcorner R_T \urcorner)$. Thus, for every k it follows that $x \neq \overline{k}$. In particular, we have (a) that $x \neq \overline{n}$. We also have $\lnot(x = \overline{0} \lor x = \overline{1} \lor \dots \lor x = \overline{n-1})$ and so by Lemma 4.23, (b) $\lnot(x < \overline{n})$. By Lemma 4.24, $\overline{n} < x$. Since $\mathbf{Q} \vdash \mathsf{Ref}_T(\overline{n}, \ulcorner R_T \urcorner)$, we have $\overline{n} < x \land \mathsf{Ref}_T(\overline{n}, \ulcorner R_T \urcorner)$, and from that $\exists z\,(z < x \land \mathsf{Ref}_T(z, \ulcorner R_T \urcorner))$. Since x was arbitrary we get, as required, that

$$\forall x\,(\mathsf{Prf}_T(x, \ulcorner R_T \urcorner) \rightarrow \exists z\,(z < x \land \mathsf{Ref}_T(z, \ulcorner R_T \urcorner))). \qquad \square$$

5.5 Comparison with Gödel's Original Paper

It is worthwhile to spend some time with Gödel's 1931 paper. The introduction sketches the ideas we have just discussed. Even if you just skim through the paper, it is easy to see what is going on at each stage: first Gödel describes the formal system P (syntax, axioms, proof rules); then he defines the primitive recursive functions and relations; then he shows that xBy is primitive recursive, and argues that the primitive recursive functions and relations are represented in **P**. He then goes on to prove the incompleteness theorem, as above. In section 3, he shows that one can take the unprovable assertion to be a sentence in the language of arithmetic. This is the origin of the β-lemma, which is what we also used to handle sequences in showing that the recursive functions are representable in **Q**. Gödel doesn't go so far to isolate a minimal set of axioms that suffice, but we now know that **Q** will do the trick. Finally, in Section 4, he sketches a proof of the second incompleteness theorem.

5.6 The Derivability Conditions for PA

Peano arithmetic, or **PA**, is the theory extending **Q** with induction axioms for all formulas. In other words, one adds to **Q** axioms of the form

$$(A(0) \land \forall x\,(A(x) \rightarrow A(x'))) \rightarrow \forall x\, A(x)$$

for every formula A. Notice that this is really a *schema*, which is to say, infinitely many axioms (and it turns out that **PA** is *not* finitely axiomatizable). But since one can effectively determine whether or not a string of symbols is an instance of an induction axiom, the set of axioms for **PA** is computable. **PA** is a much more robust theory than **Q**. For example, one can easily prove that addition and multiplication are commutative, using induction in the usual way. In fact, most finitary number-theoretic and combinatorial arguments can be carried out in **PA**.

Since **PA** is computably axiomatized, the derivability predicate $\mathrm{Prf}_{\mathbf{PA}}(x, y)$ is computable and hence represented in **Q** (and so, in **PA**). As before, we will take $\mathsf{Prf}_{\mathbf{PA}}(x, y)$ to denote the formula representing the relation. Let $\mathsf{Prov}_{\mathbf{PA}}(y)$ be the formula $\exists x\, \mathrm{Prf}_{\mathbf{PA}}(x, y)$, which, intuitively says, "y is provable from the axioms of **PA**." The reason we need a little bit more than the axioms of **Q** is we need to know that the theory we are using is strong enough to derive a few basic facts about this derivability predicate. In fact, what we need are the following facts:

P1. If $\mathbf{PA} \vdash A$, then $\mathbf{PA} \vdash \mathsf{Prov}_{\mathbf{PA}}(\ulcorner A \urcorner)$

P2. For all formulas A and B,

$$\mathbf{PA} \vdash \mathsf{Prov}_{\mathbf{PA}}(\ulcorner A \rightarrow B \urcorner) \rightarrow (\mathsf{Prov}_{\mathbf{PA}}(\ulcorner A \urcorner) \rightarrow \mathsf{Prov}_{\mathbf{PA}}(\ulcorner B \urcorner))$$

P3. For every formula A,

$$\mathbf{PA} \vdash \mathsf{Prov}_{\mathbf{PA}}(\ulcorner A \urcorner) \rightarrow \mathsf{Prov}_{\mathbf{PA}}(\ulcorner \mathsf{Prov}_{\mathbf{PA}}(\ulcorner A \urcorner) \urcorner).$$

The only way to verify that these three properties hold is to describe the formula $\mathsf{Prov}_{\mathbf{PA}}(y)$ carefully and use the axioms of **PA** to describe the relevant formal proofs. Conditions (1) and (2) are easy; it is really condition (3) that requires work. (Think about what kind of work it entails ...) Carrying out the details would be tedious and uninteresting, so here we will ask you to take it on faith that **PA** has the three properties listed above. A reasonable choice of $\mathsf{Prov}_{\mathbf{PA}}(y)$ will also satisfy

P4. If $\mathbf{PA} \vdash \mathsf{Prov}_{\mathbf{PA}}(\ulcorner A \urcorner)$, then $\mathbf{PA} \vdash A$.

But we will not need this fact.

Incidentally, Gödel was lazy in the same way we are being now. At the end of the 1931 paper, he sketches the proof of the second incompleteness theorem, and promises the details in a later paper. He never got around to it; since everyone who understood the argument believed that it could be carried out (he did not need to fill in the details.)

5.7 The Second Incompleteness Theorem

How can we express the assertion that **PA** doesn't prove its own consistency? Saying **PA** is inconsistent amounts to saying that $\mathbf{PA} \vdash 0 = 1$. So we can take the consistency statement $\mathsf{Con}_{\mathbf{PA}}$ to be the sentence $\neg\mathsf{Prov}_{\mathbf{PA}}(\ulcorner 0 = 1 \urcorner)$, and then the following theorem does the job:

Theorem 5.8. *Assuming* **PA** *is consistent, then* **PA** *does not derive* $\mathsf{Con}_{\mathbf{PA}}$.

It is important to note that the theorem depends on the particular representation of $\mathsf{Con}_{\mathbf{PA}}$ (i.e., the particular representation of $\mathsf{Prov}_{\mathbf{PA}}(y)$). All we will use is that the representation of $\mathsf{Prov}_{\mathbf{PA}}(y)$ satisfies the three derivability conditions, so the theorem generalizes to any theory with a derivability predicate having these properties.

It is informative to read Gödel's sketch of an argument, since the theorem follows like a good punch line. It goes like this. Let $G_{\mathbf{PA}}$ be the Gödel sentence that we constructed in the proof of Theorem 5.6. We have shown "If **PA** is consistent, then **PA** does not derive $G_{\mathbf{PA}}$." If we formalize this *in* **PA**, we have a proof of

$$\mathsf{Con}_{\mathbf{PA}} \to \neg\mathsf{Prov}_{\mathbf{PA}}(\ulcorner G_{\mathbf{PA}} \urcorner).$$

Now suppose **PA** derives $\mathsf{Con}_{\mathbf{PA}}$. Then it derives $\neg\mathsf{Prov}_{\mathbf{PA}}(\ulcorner G_{\mathbf{PA}} \urcorner)$. But since $G_{\mathbf{PA}}$ is a Gödel sentence, this is equivalent to $G_{\mathbf{PA}}$. So **PA** derives $G_{\mathbf{PA}}$.

But: we know that if **PA** is consistent, it doesn't derive $G_{\mathbf{PA}}$! So if **PA** is consistent, it can't derive $\mathsf{Con}_{\mathbf{PA}}$.

To make the argument more precise, we will let $G_{\mathbf{PA}}$ be the Gödel sentence for **PA** and use the derivability conditions (P1)–(P3) to show that **PA** derives $\mathsf{Con}_{\mathbf{PA}} \to G_{\mathbf{PA}}$. This will show that **PA** doesn't derive $\mathsf{Con}_{\mathbf{PA}}$. Here is a sketch of the proof, in **PA**. (For simplicity, we drop the **PA** subscripts.)

$$G \leftrightarrow \neg\mathsf{Prov}(\ulcorner G \urcorner) \tag{5.5}$$

G is a Gödel sentence

$$G \to \neg\mathsf{Prov}(\ulcorner G \urcorner) \tag{5.6}$$

from eq. (5.5)

$$G \to (\mathsf{Prov}(\ulcorner G \urcorner) \to \bot) \tag{5.7}$$

from eq. (5.6) by logic

$$\mathsf{Prov}(\ulcorner G \to (\mathsf{Prov}(\ulcorner G \urcorner) \to \bot)\urcorner) \tag{5.8}$$

by from eq. (5.7) by condition P1

$$\mathsf{Prov}(\ulcorner G \urcorner) \to \mathsf{Prov}(\ulcorner (\mathsf{Prov}(\ulcorner G \urcorner) \to \bot)\urcorner) \tag{5.9}$$

from eq. (5.8) by condition P2

$$\mathsf{Prov}(\ulcorner G \urcorner) \to (\mathsf{Prov}(\ulcorner \mathsf{Prov}(\ulcorner G \urcorner)\urcorner) \to \mathsf{Prov}(\ulcorner \bot \urcorner)) \tag{5.10}$$

from eq. (5.9) by condition P2 and logic

$$\mathsf{Prov}(\ulcorner G \urcorner) \to \mathsf{Prov}(\ulcorner \mathsf{Prov}(\ulcorner G \urcorner)\urcorner) \tag{5.11}$$

by P3

$$\mathsf{Prov}(\ulcorner G \urcorner) \to \mathsf{Prov}(\ulcorner \bot \urcorner) \tag{5.12}$$

from eq. (5.10) and eq. (5.11) by logic

$$\mathsf{Con} \to \neg\mathsf{Prov}(\ulcorner G \urcorner) \tag{5.13}$$

contraposition of eq. (5.12) and $\mathsf{Con} \equiv \neg\mathsf{Prov}(\ulcorner \bot \urcorner)$

$$\mathsf{Con} \to G$$

from eq. (5.5) and eq. (5.13) by logic

The use of logic in the above just elementary facts from propositional logic, e.g., eq. (5.7) uses $\vdash \neg A \leftrightarrow (A \to \bot)$ and eq. (5.12) uses $A \to (B \to C), A \to B \vdash A \to C$. The use of condition P2 in eq. (5.9) and eq. (5.10) relies on instances of P2, $\mathsf{Prov}(\ulcorner A \to B \urcorner) \to (\mathsf{Prov}(\ulcorner A \urcorner) \to \mathsf{Prov}(\ulcorner B \urcorner))$. In the first one, $A \equiv G$ and $B \equiv \mathsf{Prov}(\ulcorner G \urcorner) \to \bot$; in the second, $A \equiv \mathsf{Prov}(\ulcorner G \urcorner)$ and $B \equiv \bot$.

The more abstract version of the second incompleteness theorem is as follows:

Theorem 5.9. *Let* **T** *be any consistent, axiomatized theory extending* **Q** *and let* $\mathsf{Prov}_T(y)$ *be any formula satisfying derivability conditions P1–P3 for* **T**. *Then* **T** *does not derive* Con_T.

The moral of the story is that no "reasonable" consistent theory for mathematics can derive its own consistency statement. Suppose **T** is a theory of mathematics that includes **Q** and Hilbert's "finitary" reasoning (whatever that may be). Then, the whole of **T** cannot derive the consistency statement of **T**, and so, a fortiori, the finitary fragment can't derive the consistency statement of **T** either. In that sense, there cannot be a finitary consistency proof for "all of mathematics."

There is some leeway in interpreting the term "finitary," and Gödel, in the 1931 paper, grants the possibility that something we may consider "finitary" may lie outside the kinds of mathematics Hilbert wanted to formalize. But Gödel was being charitable; today, it is hard to see how we might find something that can reasonably be called finitary but is not formalizable in, say, **ZFC**.

5.8 Löb's Theorem

The Gödel sentence for a theory **T** is a fixed point of $\neg\mathsf{Prov}_T(x)$, i.e., a sentence G such that

$$\mathbf{T} \vdash \neg\mathsf{Prov}_T(\ulcorner G \urcorner) \leftrightarrow G.$$

It is not derivable, because if $\mathbf{T} \vdash G$, (a) by derivability condition (1), $\mathbf{T} \vdash \mathsf{Prov}_T(\ulcorner G \urcorner)$, and (b) $\mathbf{T} \vdash G$ together with $\mathbf{T} \vdash \neg\mathsf{Prov}_T(\ulcorner G \urcorner) \leftrightarrow G$ gives $\mathbf{T} \vdash \neg\mathsf{Prov}_T(\ulcorner G \urcorner)$, and so **T** would be inconsistent. Now it is natural to ask about the status of a fixed point of $\mathsf{Prov}_T(x)$, i.e., a sentence H such that

$$\mathbf{T} \vdash \mathsf{Prov}_T(\ulcorner H \urcorner) \leftrightarrow H.$$

If it were derivable, $\mathbf{T} \vdash \mathsf{Prov}_T(\ulcorner H \urcorner)$ by condition (1), but the same conclusion follows if we apply modus ponens to the equivalence above. Hence, we don't get that **T** is inconsistent, at least not by the same argument as in the case of the Gödel sentence. This of course does not show that **T** *does* derive H.

We can make headway on this question if we generalize it a bit. The left-to-right direction of the fixed point equivalence, $\mathsf{Prov}_T(\ulcorner H \urcorner) \to H$, is an instance of a general schema called a *reflection principle*: $\mathsf{Prov}_T(\ulcorner A \urcorner) \to A$. It is called that because it expresses, in a sense, that **T** can "reflect" about what it can derive; basically it says, "If **T** can derive A, then A is true," for any A. This is true for sound theories only, of course, and this suggests that theories will in general not derive every instance of it. So which instances can a theory (strong enough, and satisfying the derivability conditions) derive? Certainly all those where A itself is derivable. And that's it, as the next result shows.

Theorem 5.10. *Let* **T** *be an axiomatizable theory extending* **Q**, *and suppose* $\mathsf{Prov}_T(y)$ *is a formula satisfying conditions P1–P3 from section 5.7. If* **T** *derives* $\mathsf{Prov}_T(\ulcorner A \urcorner) \to A$*, then in fact* **T** *derives* A.

Put differently, if $\mathbf{T} \nvdash A$, then $\mathbf{T} \nvdash \mathsf{Prov}_T(\ulcorner A \urcorner) \to A$. This result is known as Löb's theorem.

The heuristic for the proof of Löb's theorem is a clever proof that Santa Claus exists. (If you don't like that conclusion, you are free to substitute any other conclusion you would like.) Here it is:

1. Let X be the sentence, "If X is true, then Santa Claus exists."

2. Suppose X is true.

3. Then what it says holds; i.e., we have: if X is true, then Santa Claus exists.

4. Since we are assuming X is true, we can conclude that Santa Claus exists, by modus ponens from (2) and (3).

5. We have succeeded in deriving (4), "Santa Claus exists," from the assumption (2), "X is true." By conditional proof, we have shown: "If X is true, then Santa Claus exists."

6. But this is just the sentence X. So we have shown that X is true.

7. But then, by the argument (2)–(4) above, Santa Claus exists.

A formalization of this idea, replacing "is true" with "is derivable," and "Santa Claus exists" with A, yields the proof of Löb's theorem. The trick is to apply the fixed-point lemma to the formula $\mathsf{Prov}_T(y) \to A$. The fixed point of that corresponds to the sentence X in the preceding sketch.

Proof of Theorem 5.10. Suppose A is a sentence such that $\mathbf{T}$ derives $\mathsf{Prov}_T(\ulcorner A \urcorner) \to A$. Let $B(y)$ be the formula $\mathsf{Prov}_T(y) \to A$, and use the fixed-point lemma to find a sentence D such that $\mathbf{T}$ derives $D \leftrightarrow B(\ulcorner D \urcorner)$. Then each of the following is derivable in $\mathbf{T}$:

$$D \leftrightarrow (\mathsf{Prov}_T(\ulcorner D \urcorner) \to A) \tag{5.14}$$

D is a fixed point of $B(y)$

$$D \to (\mathsf{Prov}_T(\ulcorner D \urcorner) \to A) \tag{5.15}$$

from eq. (5.14)

$$\mathsf{Prov}_T(\ulcorner D \to (\mathsf{Prov}_T(\ulcorner D \urcorner) \to A) \urcorner) \tag{5.16}$$

from eq. (5.15) by condition P1

$$\mathsf{Prov}_T(\ulcorner D \urcorner) \to \mathsf{Prov}_T(\ulcorner \mathsf{Prov}_T(\ulcorner D \urcorner) \to A \urcorner) \tag{5.17}$$

from eq. (5.16) using condition P2

$$\mathsf{Prov}_T(\ulcorner D \urcorner) \to (\mathsf{Prov}_T(\ulcorner \mathsf{Prov}_T(\ulcorner D \urcorner) \urcorner) \to \mathsf{Prov}_T(\ulcorner A \urcorner)) \tag{5.18}$$

from eq. (5.17) using P2 again

$$\mathsf{Prov}_T(\ulcorner D \urcorner) \to \mathsf{Prov}_T(\ulcorner \mathsf{Prov}_T(\ulcorner D \urcorner) \urcorner) \tag{5.19}$$

by derivability condition P3

$$\mathsf{Prov}_T(\ulcorner D \urcorner) \to \mathsf{Prov}_T(\ulcorner A \urcorner) \tag{5.20}$$

from eq. (5.18) and eq. (5.19)

$$\mathsf{Prov}_T(\ulcorner A \urcorner) \to A \tag{5.21}$$

by assumption of the theorem

$$\mathsf{Prov}_T(\ulcorner D \urcorner) \to A \tag{5.22}$$

from eq. (5.20) and eq. (5.21)

$$(\mathsf{Prov}_T(\ulcorner D \urcorner) \to A) \to D \tag{5.23}$$

from eq. (5.14)

$$D \tag{5.24}$$

from eq. (5.22) and eq. (5.23)

$$\mathsf{Prov}_T(\ulcorner D \urcorner) \tag{5.25}$$

from eq. (5.24) by condition P1

A from eq. (5.21) and eq. (5.25) □

With Löb's theorem in hand, there is a short proof of the first incompleteness theorem (for theories having a derivability predicate satisfying conditions P1–P3: if $\mathbf{T} \vdash \mathsf{Prov}_T(\ulcorner \bot \urcorner) \to \bot$, then $\mathbf{T} \vdash \bot$. If $\mathbf{T}$ is consistent, $\mathbf{T} \nvdash \bot$. So, $\mathbf{T} \nvdash \mathsf{Prov}_T(\ulcorner \bot \urcorner) \to \bot$, i.e., $\mathbf{T} \nvdash \mathsf{Con}_{\mathbf{T}}$. We can also apply it to show that H, the fixed point of $\mathsf{Prov}_T(x)$, is derivable. For since

$$\mathbf{T} \vdash \mathsf{Prov}_T(\ulcorner H \urcorner) \leftrightarrow H$$

in particular

$$\mathbf{T} \vdash \mathsf{Prov}_T(\ulcorner H \urcorner) \to H$$

and so by Löb's theorem, $\mathbf{T} \vdash H$.

5.9 The Undefinability of Truth

The notion of *definability* depends on having a formal semantics for the language of arithmetic. We have described a set of formulas and sentences in the language of arithmetic. The "intended interpretation" is to read such sentences as making assertions about the natural numbers, and such an assertion can be true or false. Let N be the structure with domain $\mathbb{N}$ and the standard interpretation for the symbols in the language of arithmetic. Then $N \vDash A$ means "A is true in the standard interpretation."

Definition 5.11. A relation $R(x_1, \dots, x_k)$ of natural numbers is *definable* in N if and only if there is a formula $A(x_1, \dots, x_k)$ in the language of arithmetic such that for every $n_1, \dots, n_k$, $R(n_1, \dots, n_k)$ if and only if $N \vDash A(\overline{n}_1, \dots, \overline{n}_k)$.

Put differently, a relation is definable in N if and only if it is representable in the theory **TA**, where $\mathbf{TA} = \{A : N \vDash A\}$ is the set of true sentences of arithmetic. (If this is not immediately clear to you, you should go back and check the definitions and convince yourself that this is the case.)

Lemma 5.12. *Every computable relation is definable in* N.

Proof. It is easy to check that the formula representing a relation in **Q** defines the same relation in N. □

Now one can ask, is the converse also true? That is, is every relation definable in N computable? The answer is no. For example:

Lemma 5.13. *The halting relation is definable in* $\mathbf{N}$.

Proof. Let H be the halting relation, i.e.,

$$H = \{\langle e, x\rangle : \exists s\, T(e, x, s)\}.$$

Let D_T define T in $\mathbf{N}$. Then

$$H = \{\langle e, x\rangle : \mathbf{N} \models \exists s\, D_T(\overline{e}, \overline{x}, s)\},$$

so $\exists s\, D_T(z, x, s)$ defines H in $\mathbf{N}$. □

What about **TA** itself? Is it definable in arithmetic? That is: is the set $\{^{\#}A^{\#} : \mathbf{N} \models A\}$ definable in arithmetic? Tarski's theorem answers this in the negative.

Theorem 5.14. *The set of true sentences of arithmetic is not definable in arithmetic.*

Proof. Suppose $D(x)$ defined it, i.e., $\mathbf{N} \models A$ iff $\mathbf{N} \models D(\ulcorner A \urcorner)$. By the fixed-point lemma, there is a formula A such that $\mathbf{Q} \vdash A \leftrightarrow \neg D(\ulcorner A \urcorner)$, and hence $\mathbf{N} \models A \leftrightarrow \neg D(\ulcorner A \urcorner)$. But then $\mathbf{N} \models A$ if and only if $\mathbf{N} \models \neg D(\ulcorner A \urcorner)$, which contradicts the fact that $D(y)$ is supposed to define the set of true statements of arithmetic. □

Tarski applied this analysis to a more general philosophical notion of truth. Given any language L, Tarski argued that an adequate notion of truth for L would have to satisfy, for each sentence X,

'X' is true if and only if X.

Tarski's oft-quoted example, for English, is the sentence

'Snow is white' is true if and only if snow is white.

However, for any language strong enough to represent the diagonal function, and any linguistic predicate $T(x)$, we can construct a sentence X satisfying "X if and only if not T('X')." Given that we do not want a truth predicate to declare some sentences to be both true and false, Tarski concluded that one cannot specify a truth predicate for all sentences in a language without, somehow, stepping outside the bounds of the language. In other words, a the truth predicate for a language cannot be defined in the language itself.

Summary

The **first incompleteness theorem** states that for any consistent, axiomatizable theory $\mathbf{T}$ that extends $\mathbf{Q}$, there is a sentence $G_{\mathbf{T}}$ such that $\mathbf{T} \nvdash G_{\mathbf{T}}$. $G_{\mathbf{T}}$ is constructed in such a way that $G_{\mathbf{T}}$, in a roundabout way, says "$\mathbf{T}$ does not prove $G_{\mathbf{T}}$." Since $\mathbf{T}$ does not, in fact, prove it, what it says is true. If $N \vDash \mathbf{T}$, then $\mathbf{T}$ does not prove any false claims, so $\mathbf{T} \nvdash \lnot G_{\mathbf{T}}$. Such a sentence is **independent** or **undecidable** in $\mathbf{T}$. Gödel's original proof established that $G_{\mathbf{T}}$ is independent on the assumption that $\mathbf{T}$ is **ω-consistent**. Rosser improved the result by finding a different sentence $R_{\mathbf{T}}$ with is neither provable nor refutable in $\mathbf{T}$ as long as ThT is simply consistent.

The construction of $G_{\mathbf{T}}$ is effective: given an axiomatization of $\mathbf{T}$ we could, in principle, write down $G_{\mathbf{T}}$. The "roundabout way" in which $G_{\mathbf{T}}$ states its own unprovability, is a special case of a general result, the **fixed-point lemma**. It states that for any formula $B(y)$ in $\mathcal{L}_A$, there is a sentence A such that $\mathbf{Q} \vdash A \leftrightarrow B(\ulcorner A \urcorner)$. (Here, $\overline{\ulcorner A \urcorner}$ is the standard numeral for the Gödel number of A, i.e., $^{\#}A^{\#}$.) To obtain $G_{\mathbf{T}}$, we use the formula $\lnot \mathsf{Prov}_{\mathbf{T}}(y)$ as $B(y)$. We get $\mathsf{Prov}_{\mathbf{T}}$ as the culmination of our previous efforts: We know that $\mathrm{Prf}_{\mathbf{T}}(n, m)$, which holds if n is the Gödel number of a derivation of the sentence with Gödel number m from $\mathbf{T}$, is primitive recursive. We also know that $\mathbf{Q}$ represents all primitive recursive relations, and so there is some formula $\mathsf{Prf}\mathbf{T}(x, y)$ that represents $\mathrm{Prf}_{\mathbf{T}}$ in $\mathbf{Q}$. The **provability predicate** for $\mathbf{T}$ is $\mathrm{Prov}_{\mathbf{T}}(y)$ is $\exists x\, \mathrm{Prf}_{ThT}(x, y)$ then expresses provability in $\mathbf{T}$. (It doesn't represent it though: if $\mathbf{T} \vdash A$, then $\mathbf{Q} \vdash \mathsf{Prov}_{\mathbf{T}}(\ulcorner A \urcorner)$; but if $\mathbf{T} \nvdash A$, then $\mathbf{Q}$ does not in general prove $\neg\mathsf{Prov}_{\mathbf{T}}(\ulcorner A \urcorner)$.)

The **second incompleteness theorem** establishes that the sentence $\mathsf{Con}_{\mathbf{T}}$ that expresses that $\mathbf{T}$ is consistent, i.e., $\mathbf{T}$ also does not prove $\neg\mathsf{Prov}_{\mathbf{T}}(\ulcorner \bot \urcorner)$. The proof of the second incompleteness theorem requires some additional conditions on $\mathbf{T}$, the **provability conditions**. **PA** satisfies them, although $\mathbf{Q}$ does not. Theories that satisfy the provability conditions also satisfy **Löb's theorem**: $\mathbf{T} \vdash \mathsf{Prov}_{\mathbf{T}}(\ulcorner A \urcorner) \rightarrow A$ iff $\mathbf{T} \vdash A$.

The fixed-point theorem also has another important consequence. We say a property Rn is **definable** in $\mathcal{L}_A$ if there is a formula $A_R(x)$ such that $\mathbf{N} \vDash A_R(\overline{n})$ iff Rn holds. For instance, $\mathrm{Prov}_{\mathbf{T}}$ is definable, since $\mathsf{Prov}_{\mathbf{T}}$ defines it. The property n has iff it is the Gödel number of a sentence true in $\mathbf{N}$, however, is not definable. This is **Tarski's theorem** about the undefinability of truth.

Problems

Problem 5.1. Every ω-consistent theory is consistent. Show that the converse does not hold, i.e., that there are consistent but ω-inconsistent theories. Do this by showing that $\mathbf{Q} \cup \{\neg G_{\mathbf{Q}}\}$ is consistent but ω-inconsistent.

Problem 5.2. Show that **PA** derives $G_{\mathbf{PA}} \rightarrow \mathsf{Con}_{\mathbf{PA}}$.

Problem 5.3. Let $\mathbf{T}$ be a computably axiomatized theory, and let Prov_T be a derivability predicate for $\mathbf{T}$. Consider the following four statements:

1. If $T \vdash A$, then $T \vdash \mathsf{Prov}_T(\ulcorner A \urcorner)$.
2. $T \vdash A \rightarrow \mathsf{Prov}_T(\ulcorner A \urcorner)$.
3. If $T \vdash \mathsf{Prov}_T(\ulcorner A \urcorner)$, then $T \vdash A$.

4. $T \vdash \mathsf{Prov}_T(\ulcorner A \urcorner) \to A$

Under what conditions are each of these statements true?

Problem 5.4. Show that $Q(n) \Leftrightarrow n \in \{{}^{\#}A^{\#} : \mathbf{Q} \vdash A\}$ is definable in arithmetic.

CHAPTER 6

Models of Arithmetic

6.1 Introduction

The *standard model* of arithmetic is the structure $\mathbf{N}$ with $|\mathbf{N}| = \mathbb{N}$ in which o, $\prime$, $+$, $\times$, and $<$ are interpreted as you would expect. That is, o is 0, $\prime$ is the successor function, $+$ is interpeted as addition and $\times$ as multiplication of the numbers in $\mathbb{N}$. Specifically,

$$\begin{aligned} \mathsf{o}^{\mathbf{N}} &= 0 \\ \prime^{\mathbf{N}}(n) &= n + 1 \\ +^{\mathbf{N}}(n, m) &= n + m \\ \times^{\mathbf{N}}(n, m) &= nm \end{aligned}$$

Of course, there are structures for $\mathcal{L}_A$ that have domains other than $\mathbb{N}$. For instance, we can take $\mathbf{M}$ with domain $|\mathbf{M}| = \{a\}^*$ (the finite sequences of the single symbol a, i.e., $\emptyset$, a, aa, aaa, ...), and interpretations

$$\begin{aligned} \mathsf{o}^{\mathbf{M}} &= \emptyset \\ \prime^{\mathbf{M}}(s) &= s \frown a \\ +^{\mathbf{M}}(n, m) &= a^{n+m} \end{aligned}$$

$$\times^M(n, m) = a^{nm}$$

These two structures are "essentially the same" in the sense that the only difference is the elements of the domains but not how the elements of the domains are related among each other by the interpretation functions. We say that the two structures are *isomorphic.*

It is an easy consequence of the compactness theorem that any theory true in N also has models that are not isomorphic to N. Such structures are called *non-standard.* The interesting thing about them is that while the elements of a standard model (i.e., N, but also all structures isomorphic to it) are exhausted by the values of the standard numerals $\overline{n}$, i.e.,

$$|N| = \{\mathrm{Val}^N(\overline{n}) : n \in \mathbb{N}\}$$

that isn't the case in non-standard models: if M is non-standard, then there is at least one $x \in |M|$ such that $x \neq \mathrm{Val}^M(\overline{n})$ for all n.

These non-standard elements are pretty neat: they are "infinite natural numbers." But their existence also explains, in a sense, the incompleteness phenomena. Consider an example, e.g., the consistency statement for Peano arithmetic, $\mathsf{Con}_{\mathbf{PA}}$, i.e., $\neg\exists x\, \mathsf{Prf}_{\mathbf{PA}}(x, \ulcorner\bot\urcorner)$. Since **PA** neither proves $\mathsf{Con}_{\mathbf{PA}}$ nor $\neg\mathsf{Con}_{\mathbf{PA}}$, either can be consistently added to **PA**. Since **PA** is consistent, $N \vDash \mathsf{Con}_{\mathbf{PA}}$, and consequently $N \nvDash \neg\mathsf{Con}_{\mathbf{PA}}$. So N is *not* a model of $\mathbf{PA} \cup \{\neg\mathsf{Con}_{\mathbf{PA}}\}$, and all its models must be nonstandard. Models of $\mathbf{PA} \cup \{\neg\mathsf{Con}_{\mathbf{PA}}\}$ must contain some element that serves as the witness that makes $\exists x\, \mathsf{Prf}_{\mathbf{PA}}(\ulcorner\bot\urcorner)$ true, i.e., a Gödel number of a derivation of a contradiction from **PA**. Such an element can't be standard—since $\mathbf{PA} \vdash \neg\mathsf{Prf}_{\mathbf{PA}}(\overline{n}, \ulcorner\bot\urcorner)$ for every n.

6.2 Reducts and Expansions

Often it is useful or necessary to compare languages which have symbols in common, as well as structures for these languages. The most comon case is when all the symbols in a language $\mathcal{L}$

are also part of a language $\mathcal{L}'$, i.e., $\mathcal{L} \subseteq \mathcal{L}'$. An $\mathcal{L}$-structure M can then always be expanded to an $\mathcal{L}'$-structure by adding interpretations of the additional symbols while leaving the interpretations of the common symbols the same. On the other hand, from an $\mathcal{L}'$-structure M' we can obtain an $\mathcal{L}$-structure simpy by "forgetting" the interpretations of the symbols that do not occur in $\mathcal{L}$.

Definition 6.1. Suppose $\mathcal{L} \subseteq \mathcal{L}'$, M is an $\mathcal{L}$-structure and M' is an $\mathcal{L}'$-structure. M is the *reduct* of M' to $\mathcal{L}$, and M' is an *expansion* of M to $\mathcal{L}'$ iff

1. $|M| = |M'|$
2. For every constant symbol $c \in \mathcal{L}$, $c^M = c^{M'}$.
3. For every function symbol $f \in \mathcal{L}$, $f^M = f^{M'}$.
4. For every predicate symbol $P \in \mathcal{L}$, $P^M = P^{M'}$.

Proposition 6.2. *If an $\mathcal{L}$-structure M is a reduct of an $\mathcal{L}'$-structure M', then for all $\mathcal{L}$-sentences A,*

$$M \vDash A \text{ iff } M' \vDash A.$$

Proof. Exercise. □

Definition 6.3. When we have an $\mathcal{L}$-structure M, and $\mathcal{L}' = \mathcal{L} \cup \{P\}$ is the expansion of $\mathcal{L}$ obtained by adding a single n-place predicate symbol P, and $R \subseteq |M|^n$ is an n-place relation, then we write (M, R) for the expansion M' of M with $P^{M'} = R$.

6.3 Isomorphic Structures

First-order structures can be alike in one of two ways. One way in which the can be alike is that they make the same sentences true. We call such structures *elementarily equivalent.* But structures can be very different and still make the same sentences true—for instance, one can be countable and the other not. This is because there are lots of features of a structure that cannot be expressed in first-order languages, either because the language is not rich enough, or because of fundamental limitations of first-order logic such as the Löwenheim-Skolem theorem. So another, stricter, aspect in which structures can be alike is if they are fundamentally the same, in the sense that they only differ in the objects that make them up, but not in their structural features. A way of making this precise is by the notion of an *isomorphism.*

Definition 6.4. Given two structures $\mathbf{M}$ and $\mathbf{M}'$ for the same language $\mathcal{L}$, we say that $\mathbf{M}$ is *elementarily equivalent to* $\mathbf{M}'$, written $\mathbf{M} \equiv \mathbf{M}'$, if and only if for every sentence A of $\mathcal{L}$, $\mathbf{M} \vDash A$ iff $\mathbf{M}' \vDash A$.

Definition 6.5. Given two structures $\mathbf{M}$ and $\mathbf{M}'$ for the same language $\mathcal{L}$, we say that $\mathbf{M}$ is *isomorphic to* $\mathbf{M}'$, written $\mathbf{M} \simeq \mathbf{M}'$, if and only if there is a function $h\colon |\mathbf{M}| \to |\mathbf{M}'|$ such that:

1. h is injective: if $h(x) = h(y)$ then $x = y$;
2. h is surjective: for every $y \in |\mathbf{M}'|$ there is $x \in |\mathbf{M}|$ such that $h(x) = y$;
3. for every constant symbol c: $h(c^{\mathbf{M}}) = c^{\mathbf{M}'}$;
4. for every n-place predicate symbol P:
$$\langle a_1, \dots, a_n \rangle \in P^{\mathbf{M}} \quad \text{iff} \quad \langle h(a_1), \dots, h(a_n) \rangle \in P^{\mathbf{M}'};$$
5. for every n-place function symbol f:
$$h(f^{\mathbf{M}}(a_1, \dots, a_n)) = f^{\mathbf{M}'}(h(a_1), \dots, h(a_n)).$$

Theorem 6.6. *If $M \simeq M'$ then $M \equiv M'$.*

Proof. Let h be an isomorphism of M onto M'. For any assignment s, $h \circ s$ is the composition of h and s, i.e., the assignment in M' such that $(h \circ s)(x) = h(s(x))$. By induction on t and A one can prove the stronger claims:

a. $h(\mathrm{Val}_s^M(t)) = \mathrm{Val}_{h\circ s}^{M'}(t)$.

b. $M, s \models A$ iff $M', h \circ s \models A$.

The first is proved by induction on the complexity of t.

1. If $t \equiv c$, then $\mathrm{Val}_s^M(c) = c^M$ and $\mathrm{Val}_{h\circ s}^{M'}(c) = c^{M'}$. Thus, $h(\mathrm{Val}_s^M(t)) = h(c^M) = c^{M'}$ (by (3) of Definition 6.5) $= \mathrm{Val}_{h\circ s}^{M'}(t)$.

2. If $t \equiv x$, then $\mathrm{Val}_s^M(x) = s(x)$ and $\mathrm{Val}_{h\circ s}^{M'}(x) = h(s(x))$. Thus, $h(\mathrm{Val}_s^M(x)) = h(s(x)) = \mathrm{Val}_{h\circ s}^{M'}(x)$.

3. If $t \equiv f(t_1, \dots, t_n)$, then

$$\mathrm{Val}_s^M(t) = f^M(\mathrm{Val}_s^M(t_1), \dots, \mathrm{Val}_s^M(t_n)) \quad \text{and}$$
$$\mathrm{Val}_{h\circ s}^{M'}(t) = f^M(\mathrm{Val}_{h\circ s}^{M'}(t_1), \dots, \mathrm{Val}_{h\circ s}^{M'}(t_n)).$$

The induction hypothesis is that for each i, $h(\mathrm{Val}_s^M(t_i)) = \mathrm{Val}_{h\circ s}^{M'}(t_i)$. So,

$$\begin{aligned} h(\mathrm{Val}_s^M(t)) &= h(f^M(\mathrm{Val}_s^M(t_1), \dots, \mathrm{Val}_s^M(t_n)) \\ &= h(f^M(\mathrm{Val}_{h\circ s}^{M'}(t_1), \dots, \mathrm{Val}_{h\circ s}^{M'}(t_n)) && (6.1) \\ &= f^{M'}(\mathrm{Val}_{h\circ s}^{M'}(t_1), \dots, \mathrm{Val}_{h\circ s}^{M'}(t_n)) && (6.2) \\ &= \mathrm{Val}_{h\circ s}^{M'}(t) \end{aligned}$$

Here, eq. (6.1) follows by induction hypothesis and eq. (6.2) by (5) of Definition 6.5.

Part (2) is left as an exercise.

If A is a sentence, the assignments s and $h \circ s$ are irrelevant, and we have $M \models A$ iff $M' \models A$. □

Definition 6.7. An *automorphism* of a structure $\mathfrak{M}$ is an isomorphism of $\mathfrak{M}$ onto itself.

6.4 The Theory of a Structure

Every structure M makes some sentences true, and some false. The set of all the sentences it makes true is called its *theory*. That set is in fact a theory, since anything it entails must be true in all its models, including M.

Definition 6.8. Given a structure M, the *theory* of M is the set $\mathrm{Th}(M)$ of sentences that are true in M, i.e., $\mathrm{Th}(M) = \{A : M \vDash A\}$.

We also use the term "theory" informally to refer to sets of sentences having an intended interpretation, whether deductively closed or not.

Proposition 6.9. *For any* M, $\mathrm{Th}(M)$ *is complete.*

Proof. For any sentence A either $M \vDash A$ or $M \vDash \neg A$, so either $A \in \mathrm{Th}(M)$ or $\neg A \in \mathrm{Th}(M)$. □

Proposition 6.10. *If* $N \vDash A$ *for every* $A \in \mathrm{Th}(M)$, *then* $M \equiv N$.

Proof. Since $N \vDash A$ for all $A \in \mathrm{Th}(M)$, $\mathrm{Th}(M) \subseteq \mathrm{Th}(N)$. If $N \vDash A$, then $N \nvDash \neg A$, so $\neg A \notin \mathrm{Th}(M)$. Since $\mathrm{Th}(M)$ is complete, $A \in \mathrm{Th}(M)$. So, $\mathrm{Th}(N) \subseteq \mathrm{Th}(M)$, and we have $M \equiv N$. □

Remark 1. Consider $R = \langle \mathbb{R}, < \rangle$, the structure whose domain is the set $\mathbb{R}$ of the real numbers, in the language comprising only a 2-place predicate symbol interpreted as the $<$ relation over the reals. Clearly R is uncountable; however, since $\mathrm{Th}(R)$ is obviously consistent, by the Löwenheim-Skolem theorem it has a countable model, say S, and by Proposition 6.10, $R \equiv S$. Moreover, since R and S are not isomorphic, this shows that the converse of Theorem 6.6 fails in general.

6.5 Standard Models of Arithmetic

The language of arithmetic $\mathcal{L}_A$ is obviously intended to be about numbers, specifically, about natural numbers. So, "the" standard model N is special: it is the model we want to talk about. But in logic, we are often just interested in structural properties, and any two structures taht are isomorphic share those. So we can be a bit more liberal, and consider any structure that is isomorphic to N "standard."

Definition 6.11. A structure for $\mathcal{L}_A$ is *standard* if it is isomorphic to N.

Proposition 6.12. *If a structure M standard, its domain is the set of values of the standard numerals, i.e.,*

$$|M| = \{\mathrm{Val}^M(\overline{n}) : n \in \mathbb{N}\}$$

Proof. Clearly, every $\mathrm{Val}^M(\overline{n}) \in |M|$. We just have to show that every $x \in |M|$ is equal to $\mathrm{Val}^M(\overline{n})$ for some n. Since M is standard, it is isomorphic to N. Suppose $g\colon \mathbb{N} \to |M|$ is an isomorphism. Then $g(n) = g(\mathrm{Val}^N(\overline{n})) = \mathrm{Val}^M(\overline{n})$. But for every $x \in |M|$, there is an $n \in \mathbb{N}$ such that $g(n) = x$, since g is surjective. □

If a structure M for $\mathcal{L}_A$ is standard, the elements of its domain can all be named by the standard numerals $\overline{0}, \overline{1}, \overline{2}, \ldots$, i.e., the terms o, o', o'', etc. Of course, this does not mean that the elements of $|M|$ *are* the numbers, just that we can pick them out the same way we can pick out the numbers in $|N|$.

Proposition 6.13. *If $M \vDash \mathbf{Q}$, and $|M| = \{\mathrm{Val}^M(\overline{n}) : n \in \mathbb{N}\}$, then M is standard.*

Proof. We have to show that M is isomorphic to N. Consider the function $g\colon \mathbb{N} \to |M|$ defined by $g(n) = \mathrm{Val}^M(\overline{n})$. By the hypothesis, g is surjective. It is also injective: $\mathbf{Q} \vdash \overline{n} \neq \overline{m}$ whenever

$n \neq m$. Thus, since $M \vDash \mathbf{Q}$, $M \vDash \overline{n} \neq \overline{m}$, whenever $n \neq m$. Thus, if $n \neq m$, then $\mathrm{Val}^M(\overline{n}) \neq \mathrm{Val}^M(\overline{m})$, i.e., $g(n) \neq g(m)$.

We also have to verify that g is an isomorphism.

1. We have $g(\mathrm{o}^N) = g(0)$ since, $\mathrm{o}^N = 0$. By definition of g, $g(0) = \mathrm{Val}^M(\overline{0})$. But $\overline{0}$ is just o, and the value of a term which happens to be a constant symbol is given by what the structure assigns to that constant symbol, i.e., $\mathrm{Val}^M(\mathrm{o}) = \mathrm{o}^M$. So we have $g(\mathrm{o}^N) = \mathrm{o}^M$ as required.

2. $g({\prime}^N(n)) = g(n+1)$, since $\prime$ in N is the successor function on $\mathbb{N}$. Then, $g(n+1) = \mathrm{Val}^M(\overline{n+1})$ by definition of g. But $\overline{n+1}$ is the same term as $\overline{n}'$, so $\mathrm{Val}^M(\overline{n+1}) = \mathrm{Val}^M(\overline{n}')$. By the definition of the value function, this is $= {\prime}^M(\mathrm{Val}^M(\overline{n}))$. Since $\mathrm{Val}^M(\overline{n}) = g(n)$ we get $g({\prime}^N(n)) = {\prime}^M(g(n))$.

3. $g(+^N(n,m)) = g(n+m)$, since $+$ in N is the addition function on $\mathbb{N}$. Then, $g(n+m) = \mathrm{Val}^M(\overline{n+m})$ by definition of g. But $\mathbf{Q} \vdash \overline{n+m} = (\overline{n} + \overline{m})$, so $\mathrm{Val}^M(\overline{n+m}) = \mathrm{Val}^M(\overline{n} + \overline{m})$. By the definition of the value function, this is $= +^M(\mathrm{Val}^M(\overline{n}), \mathrm{Val}^M(\overline{m}))$. Since $\mathrm{Val}^M(\overline{n}) = g(n)$ and $\mathrm{Val}^M(\overline{m}) = g(m)$, we get $g(+^N(n,m)) = +^M(g(n), g(m))$.

4. $g(\times^N(n,m)) = \times^M(g(n), g(m))$: Exercise.

5. $\langle n, m\rangle \in <^N$ iff $n < m$. If $n < m$, then $\mathbf{Q} \vdash \overline{n} < \overline{m}$, and also $M \vDash \overline{n} < \overline{m}$. Thus $\langle \mathrm{Val}^M(\overline{n}), \mathrm{Val}^M(\overline{m})\rangle \in <^M$, i.e., $\langle g(n), g(m)\rangle \in <^M$. If $n \not< m$, then $\mathbf{Q} \vdash \neg\overline{n} < \overline{m}$, and consequently $M \nvDash \overline{n} < \overline{m}$. Thus, as before, $\langle g(n), g(m)\rangle \notin <^M$. Together, we get: $\langle n, m\rangle \in <^N$ iff $\langle g(n), g(m)\rangle \in <^M$. □

The function g is the most obvious way of defining a mapping from $\mathbb{N}$ to the domain of any other structure M for $\mathcal{L}_A$, since every such M contains elements named by $\overline{0}$, $\overline{1}$, $\overline{2}$, etc. So it isn't surprising that if M makes at least some basic statements about the $\overline{n}$'s true in the same way that N does, and g is also bijective, then g will turn into an isomorphism. In fact, if $|M|$ contains no elements other than what the $\overline{n}$'s name, it's the only one.

Proposition 6.14. *If $\mathbf{M}$ is standard, then g from the proof of Proposition 6.13 is the only isomorphism from $\mathbf{N}$ to $\mathbf{M}$.*

Proof. Suppose $h \colon \mathbb{N} \to |\mathbf{M}|$ is an isomorphism between $\mathbf{N}$ and $\mathbf{M}$. We show that $g = h$ by induction on n. If $n = 0$, then $g(0) = \mathrm{o}^{\mathbf{M}}$ by definition of g. But since h is an isomorphism, $h(0) = h(\mathrm{o}^{\mathbf{N}}) = \mathrm{o}^{\mathbf{M}}$, so $g(0) = h(0)$.

Now consider the case for $n + 1$. We have

$$\begin{aligned} g(n+1) &= \mathrm{Val}^{\mathbf{M}}(\overline{n+1}) \text{ by definition of } g \\ &= \mathrm{Val}^{\mathbf{M}}(\overline{n}') \text{ since } \overline{n+1} \equiv \overline{n}' \\ &= {\prime}^{\mathbf{M}}(\mathrm{Val}^{\mathbf{M}}(\overline{n})) \text{ by definition of } \mathrm{Val}^{\mathbf{M}}(t') \\ &= {\prime}^{\mathbf{M}}(g(n)) \text{ by definition of } g \\ &= {\prime}^{\mathbf{M}}(h(n)) \text{ by induction hypothesis} \\ &= h({\prime}^{\mathbf{N}}(n)) \text{ since } h \text{ is an isomorphism} \\ &= h(n+1) \end{aligned}$$

□

For any countably infinite set M, there's a bijection between $\mathbb{N}$ and M, so every such set M is potentially the domain of a standard model $\mathbf{M}$. In fact, once you pick an object $z \in M$ and a suitable function s as $\mathrm{o}^{\mathbf{M}}$ and ${\prime}^{\mathbf{M}}$, the interpretations of $+$, $\times$, and $<$ is already fixed. Only functions $s \colon M \to M \setminus \{z\}$ that are both injective and surjective are suitable in a standard model as ${\prime}^{\mathbf{M}}$. The range of s cannot contain z, since otherwise $\forall x\, \mathrm{o} \neq x'$ would be false. That sentence is true in $\mathbf{N}$, and so $\mathbf{M}$ also has to make it true. The function s has to be injective, since the successor function ${\prime}^{\mathbf{N}}$ in $\mathbf{N}$ is, and that ${\prime}^{\mathbf{N}}$ is injective is expressed by a sentence true in $\mathbf{N}$. It has to be surjective because otherwise there would be some $x \in M \setminus \{z\}$ not in the domain of s, i.e., the sentence $\forall x\, (x = \mathrm{o} \lor \exists y\, y' = x)$ would be false in $\mathbf{M}$—but it is true in $\mathbf{N}$.

6.6 Non-Standard Models

We call a structure for $\mathcal{L}_A$ standard if it is isomorphic to $\mathfrak{N}$. If a structure isn't isomorphic to $\mathfrak{N}$, it is called non-standard.

Definition 6.15. A structure $\mathfrak{M}$ for $\mathcal{L}_A$ is *non-standard* if it is not isomorphic to $\mathfrak{N}$. The elements $x \in |\mathfrak{M}|$ which are equal to $\mathrm{Val}^{\mathfrak{M}}(\overline{n})$ for some $n \in \mathbb{N}$ are called *standard numbers* (of $\mathfrak{M}$), and those not, *non-standard numbers.*

By Proposition 6.12, any standard structure for $\mathcal{L}_A$ contains only standard elements. Consequently, a non-standard structure must contain at least one non-standard element. In fact, the existence of a non-standard element guarantees that the structure is non-standard.

Proposition 6.16. *If a structure $\mathfrak{M}$ for $\mathcal{L}_A$ contains a non-standard number, $\mathfrak{M}$ is non-standard.*

Proof. Suppose not, i.e., suppose $\mathfrak{M}$ standard but contains a non-standard number x. Let $g\colon \mathbb{N} \to |\mathfrak{M}|$ be an isomorphism. It is easy to see (by induction on n) that $g(\mathrm{Val}^{\mathfrak{N}}(\overline{n})) = \mathrm{Val}^{\mathfrak{M}}(\overline{n})$. In other words, g maps standard numbers of $\mathfrak{N}$ to standard numbers of $\mathfrak{M}$. If $\mathfrak{M}$ contains a non-standard number, g cannot be surjective, contrary to hypothesis. □

It is easy enough to specify non-standard structures for $\mathcal{L}_A$. For instance, take the structure with domain $\mathbb{Z}$ and interpret all non-logical symbols as usual. Since negative numbers are not values of $\overline{n}$ for any n, this structure is non-standard. Of course, it will not be a *model* of arithmetic in the sense that it makes the same sentences true as $\mathfrak{N}$. For instance, $\forall x\, x' \neq 0$ is false. However, we can prove that non-standard models of arithmetic exist easily enough, using the compactness theorem.

Proposition 6.17. *Let* **TA** $= \{A : N \vDash A\}$ *be the theory of* N. **TA** *has a countable non-standard model.*

Proof. Expand $\mathcal{L}_A$ by a new constant symbol c and consider the set of sentences

$$\Gamma = \mathbf{TA} \cup \{c \neq \overline{0}, c \neq \overline{1}, c \neq \overline{2}, \dots\}$$

Any model M^c of Γ would contain an element $x = c^M$ which is non-standard, since $x \neq \mathrm{Val}^M(\overline{n})$ for all $n \in \mathbb{N}$. Also, obviously, $M^c \vDash$ **TA**, since **TA** $\subseteq \Gamma$. If we turn M^c into a structure M for $\mathcal{L}_A$ simply by forgetting about c, its domain still contains the non-standard x, and also $M \vDash$ **TA**. The latter is guaranteed since c does not occur in **TA**. So, it suffices to show that Γ has a model.

We use the compactness theorem to show that Γ has a model. If every finite subset of Γ is satisfiable, so is Γ. Consider any finite subset $\Gamma_0 \subseteq \Gamma$. Γ_0 includes some sentences of **TA** and some of the form $c \neq \overline{n}$, but only finitely many. Suppose k is the largest number so that $c \neq \overline{k} \in \Gamma_0$. Define N_k by expanding N to include the interpretation $c^{N_k} = k + 1$. $N_k \vDash \Gamma_0$: if $A \in$ **TA**, $N_k \vDash A$ since N_k is just like N in all respects except c, and c does not occur in A. And $N_k \vDash c \neq \overline{n}$, since $n \leq k$, and $\mathrm{Val}^{N_k}(c) = k + 1$. Thus, every finite subset of Γ is satisfiable. □

6.7 Models of Q

We know that there are non-standard structures that make the same sentences true as N does, i.e., is a model of **TA**. Since $N \vDash$ **Q**, any model of **TA** is also a model of **Q**. **Q** is much weaker than **TA**, e.g., **Q** $\nvdash \forall x\, \forall y\, (x+y) = (y+x)$. Weaker theories are easier to satisfy: they have more models. E.g., **Q** has models which make $\forall x\, \forall y\, (x + y) = (y + x)$ false, but those cannot also be models of **TA**, or **PA** for that matter. Models of **Q** are also relatively simple: we can specify them explicitly.

Example 6.18. Consider the structure K with domain $|K| = \mathbb{N} \cup \{a\}$ and interpretations

$$\mathrm{o}^K = 0$$

$$\prime^K(x) = \begin{cases} x+1 & \text{if } x \in \mathbb{N} \\ a & \text{if } x = a \end{cases}$$

$$+^K(x, y) = \begin{cases} x+y & \text{if } x, y \in \mathbb{N} \\ a & \text{otherwise} \end{cases}$$

$$\times^K(x, y) = \begin{cases} xy & \text{if } x, y \in \mathbb{N} \\ a & \text{otherwise} \end{cases}$$

$$<^K = \{\langle x, y\rangle : x, y \in \mathbb{N} \text{ and } x < y\} \cup \{\langle x, a\rangle : x \in |K|\}$$

To show that $K \vDash \mathbf{Q}$ we have to verify that all axioms of $\mathbf{Q}$ are true in K. For convenience, let's write x^* for $\prime^K(x)$ (the "successor" of x in K), $x \oplus y$ for $+^K(x, y)$ (the "sum" of x and y in K, $x \otimes y$ for $\times^K(x, y)$ (the "product" of x and y in K), and $x \circledless y$ for $\langle x, y\rangle \in <^K$. With these abbreviations, we can give the operations in K more perspicuously as

x	x^*
n	$n+1$
a	a

$x \oplus y$	m	a
n	$n+m$	a
a	a	a

$x \otimes y$	m	a
n	nm	a
a	a	a

We have $n \circledless m$ iff $n < m$ for $n, m \in \mathbb{N}$ and $x \circledless a$ for all $x \in |K|$.

$K \vDash \forall x \forall y\, (x' = y' \rightarrow x = y)$ since $*$ is injective. $K \vDash \forall x\, \mathrm{o} \neq x'$ since 0 is not a $*$-successor in K. $K \vDash \forall x\, (x = \mathrm{o} \lor \exists y\, x = y')$ since for every $n > 0$, $n = (n-1)^*$, and $a = a^*$.

$K \vDash \forall x\, (x + \mathrm{o}) = x$ since $n \oplus 0 = n + 0 = n$, and $a \oplus 0 = a$ by definition of $\oplus$. $K \vDash \forall x \forall y\, (x + y') = (x + y)'$ is a bit trickier. If n, m are both standard, we have:

$$(n \oplus m^*) = (n + (m+1)) = (n+m) + 1 = (n \oplus m)^*$$

since $\oplus$ and * agree with $+$ and $\prime$ on standard numbers. Now suppose $x \in |K|$. Then

$$(x \oplus a^*) = (x \oplus a) = a = a^* = (x \oplus a)^*$$

The remaining case is if $y \in |K|$ but $x = a$. Here we also have to distinguish cases according to whether $y = n$ is standard or $y = b$:

$$(a \oplus n^*) = (a \oplus (n+1)) = a = a^* = (x \oplus n)^*$$
$$(a \oplus a^*) = (a \oplus a) = a = a^* = (x \oplus a)^*$$

This is of course a bit more detailed than needed. For instance, since $a \oplus z = a$ whatever z is, we can immediately conclude $a \oplus a^* = a$. The remaining axioms can be verified the same way.

K is thus a model of $\mathbf{Q}$. Its "addition" $\oplus$ is also commutative. But there are other sentences true in N but false in K, and vice versa. For instance, $a \otimes a$, so $K \vDash \exists x\, x < x$ and $K \nvDash \forall x\, \neg x < x$. This shows that $\mathbf{Q} \nvDash \forall x\, \neg x < x$.

Example 6.19. Consider the structure L with domain $|L| = \mathbb{N} \cup \{a, b\}$ and interpretations $\prime^L = *$, $+^L = \oplus$ given by

x	x^*
n	$n+1$
a	a
b	b

$x \oplus y$	m	a	b
n	$n+m$	b	a
a	a	b	a
b	b	b	a

Since $*$ is injective, 0 is not in its range, and every $x \in |L|$ other than 0 is, axioms Q_1–Q_3 are true in L. For any x, $x \oplus 0 = x$, so Q_4 is true as well. For Q_5, consider $x \oplus y^*$ and $(x \oplus y)^*$. They are equal if x and y are both standard, since then $*$ and $\oplus$ agree with $\prime$ and $+$. If x is non-standard, and y is standard, we have $x \oplus y^* = x = x^* = (x \oplus y)^*$. If x and y are both non-standard, we have four cases:

$$a \oplus a^* = b = b^* = (a \oplus a)^*$$
$$b \oplus b^* = a = a^* = (b \oplus b)^*$$
$$b \oplus a^* = b = b^* = (b \oplus y)^*$$
$$a \oplus b^* = a = a^* = (a \oplus b)^*$$

If x is standard, but y is non-standard, we have

$$n \oplus a^* = n \oplus a = b = b^* = (n \oplus a)^*$$

$$n \oplus b^* = n \oplus b = a = a^* = (n \oplus b)^*$$

So, $L \vDash Q_5$. However, $a \oplus 0 \neq 0 \oplus a$, so $L \nvDash \forall x \, \forall y \, (x + y) = (y + x)$.

We've explicitly constructed models of **Q** in which the non-standard elements live "beyond" the standard elements. In fact, that much is required by the axioms. A non-standard element x cannot be $\otimes 0$, since $\mathbf{Q} \vdash \forall x \, \neg x < 0$ (see Lemma 4.22). Also, for every n, $\mathbf{Q} \vdash \forall x \, (x < \overline{n}' \rightarrow (x = \overline{0} \lor x = \overline{1} \lor \cdots \lor x = \overline{n}))$ (Lemma 4.23), so we can't have $a \otimes n$ for any $n > 0$.

6.8 Models of PA

Any non-standard model of **TA** is also one of **PA**. We know that non-standard models of **TA** and hence of **PA** exist. We also know that such non-standard models contain non-standard "numbers," i.e., elements of the domain that are "beyond" all the standard "numbers." But how are they arranged? How many are there? We've seen that models of the weaker theory **Q** can contain as few as a single non-standard number. But these simple structures are not models of **PA** or **TA**.

The key to understanding the structure of models of **PA** or **TA** is to see what facts are derivable in these theories. For instance, already **PA** proves that $\forall x \, x \neq x'$ and $\forall x \, \forall y \, (x+y) = (y+x)$, so this rules out simple structures (in which these sentences are false) as models of **PA**.

Suppose M is a model of **PA**. Then if $\mathbf{PA} \vdash A$, $M \vDash A$. Let's again use $\mathbf{z}$ for 0^M, $*$ for $\prime^M$, $\oplus$ for $+^M$, $\otimes$ for $\times^M$, and $\otimes$ for $<^M$. Any sentence A then states some condition about $\mathbf{z}$, $*$, $\oplus$, $\otimes$, and $\otimes$, and if $M \vDash A$ that condition must be satisfied. For instance, if $M \vDash Q_1$, i.e., $M \vDash \forall x \, \forall y \, (x' = y' \rightarrow x = y)$, then $*$ must be injective.

Proposition 6.20. *In M, $\otimes$ is a linear strict order, i.e., it satisfies:*

1. *Not $x \otimes x$ for any $x \in |M|$.*

2. *If* $x \lessdot y$ *and* $y \lessdot z$ *then* $x \lessdot z$.

3. *For any* $x \neq y$, $x \lessdot y$ *or* $y \lessdot x$

Proof. **PA** proves:

1. $\forall x\, \neg x < x$

2. $\forall x\, \forall y\, \forall z\, ((x < y \land y < z) \rightarrow x < z)$

3. $\forall x\, \forall y\, ((x < y \lor y < x) \lor x = y))$ □

Proposition 6.21. **z** *is the least element of* $|\mathbf{M}|$ *in the* $\lessdot$*-ordering. For any* x, $x \lessdot x^*$, *and* x^* *is the* $\lessdot$*-least element with that property. For any* x, *there is a unique* y *such that* $y^* = x$. *(We call* y *the "predecessor" of* x *in* $\mathbf{M}$, *and denote it by* ${}^*x$.*)*

Proof. Exercise. □

Proposition 6.22. *All standard elements of* $\mathbf{M}$ *are less than (according to* $\lessdot$*) all non-standard elements.*

Proof. We'll use n as short for $\mathrm{Val}^{\mathbf{M}}(\overline{n})$, a standard element of $\mathbf{M}$. Already **Q** proves that, for any $n \in \mathbb{N}$, $\forall x\, (x < \overline{n}' \rightarrow (x = \overline{0} \lor x = \overline{1} \lor \cdots \lor x = \overline{n}))$. There are no elements that are $\lessdot\mathbf{z}$. So if n is standard and x is non-standard, we cannot have $x \lessdot n$. By definition, a non-standard element is one that isn't $\mathrm{Val}^{\mathbf{M}}(\overline{n})$ for any $n \in \mathbb{N}$, so $x \neq n$ as well. Since $\lessdot$ is a linear order, we must have $n \lessdot x$. □

Proposition 6.23. *Every nonstandard element* x *of* $|\mathbf{M}|$ *is an element of the subset*

$$\ldots {}^{***}x \lessdot {}^{**}x \lessdot {}^{*}x \lessdot x \lessdot x^{*} \lessdot x^{**} \lessdot x^{***} \lessdot \ldots$$

We call this subset the block of x *and write it as* $[x]$. *It has no least and* *no greatest element. It can be characterized as the set of those $y \in |M|$ such that, for some standard n, $x \oplus n = y$ or $y \oplus n = x$.*

Proof. Clearly, such a set $[x]$ always exists since every element y of $|M|$ has a unique successor y^* and unique predecessor ${}^*y$. For successive elements y, y^* we have $y \otimes y^*$ and y^* is the $\otimes$-least element of $|M|$ such that y is $\otimes$-less than it. Since always ${}^*y \otimes y$ and $y \otimes y^*$, $[x]$ has no least or greatest element. If $y \in [x]$ then $x \in [y]$, for then either $y^{*\cdots*} = x$ or $x^{*\cdots*} = y$. If $y^{*\cdots*} = x$ (with n $*$'s), then $y \oplus n = x$ and conversely, since $\mathbf{PA} \vdash \forall x\, x'^{\cdots\prime} = (x + \overline{n})$ (if n is the number of $\prime$'s). □

Proposition 6.24. *If $[x] \neq [y]$ and $x \otimes y$, then for any $u \in [x]$ and any $v \in [y]$, $u \otimes v$.*

Proof. Note that $\mathbf{PA} \vdash \forall x\, \forall y\, (x < y \rightarrow (x' < y \lor x' = y))$. Thus, if $u \otimes v$, we also have $u \oplus n^* \otimes v$ for any n if $[u] \neq [v]$.

Any $u \in [x]$ is $\otimes y$: $x \otimes y$ by assumption. If $u \otimes x$, $u \otimes y$ by transitivity. And if $x \otimes u$ but $u \in [x]$, we have $u = x \oplus n^*$ for some n, and so $u \otimes y$ by the fact just proved.

Now suppose that $v \in [y]$ is $\otimes y$, i.e., $v \oplus m^* = y$ for some standard m. This rules out $v \otimes x$, otherwise $y = v \oplus m^* \otimes x$. Clearly also, $x \neq v$, otherwise $x \oplus m^* = v \oplus m^* = y$ and we would have $[x] = [y]$. So, $x \otimes v$. But then also $x \oplus n^* \otimes v$ for any n. Hence, if $x \otimes u$ and $u \in [x]$, we have $u \otimes v$. If $u \otimes x$ then $u \otimes v$ by transitivity.

Lastly, if $y \otimes v$, $u \otimes v$ since, as we've shown, $u \otimes y$ and $y \otimes v$. □

Corollary 6.25. *If $[x] \neq [y]$, $[x] \cap [y] = \emptyset$.*

Proof. Suppose $z \in [x]$ and $x \otimes y$. Then $z \otimes u$ for all $u \in [y]$. If $z \in [y]$, we would have $z \otimes z$. Similarly if $y \otimes x$. □

This means that the blocks themselves can be ordered in a way that respects $\otimes$: $[x] \otimes [y]$ iff $x \otimes y$, or, equivalently, if $u \otimes v$ for any $u \in [x]$ and $v \in [y]$. Clearly, the standard block $[0]$ is the least block. It intersects with no non-standard block, and no two non-standard blocks intersect either. Specifically, you cannot "reach" a different block by taking repeated successors or predecessors.

Proposition 6.26. *If x and y are non-standard, then $x \otimes x \oplus y$ and $x \oplus y \notin [x]$.*

Proof. If y is nonstandard, then $y \neq \mathbf{z}$. $\mathbf{PA} \vdash \forall x\,(y \neq 0 \rightarrow x < (x + y))$. Now suppose $x \oplus y \in [x]$. Since $x \otimes x \oplus y$, we would have $x \oplus n^* = x \oplus y$. But $\mathbf{PA} \vdash \forall x\, \forall y\, \forall z\, ((x + y) = (x + z) \rightarrow y = z)$ (the cancellation law for addition). This would mean $y = n^*$ for some standard n; but y is assumed to be non-standard. □

Proposition 6.27. *There is no least non-standard block.*

Proof. $\mathbf{PA} \vdash \forall x\, \exists y\, ((y + y) = x \lor (y + y)' = x)$, i.e., that every x is divisible by 2 (possibly with remainder 1). If x is non-standard, so is y. By the preceding proposition, $y \otimes y \oplus y$ and $y \oplus y \notin [y]$. Then also $y \otimes (y \oplus y)^*$ and $(y \oplus y)^* \notin [y]$. But $x = y \oplus y$ or $x = (y \oplus y)^*$, so $y \otimes x$ and $y \notin [x]$. □

Proposition 6.28. *There is no largest block.*

Proof. Exercise. □

Proposition 6.29. *The ordering of the blocks is dense. That is, if $x \otimes y$ and $[x] \neq [y]$, then there is a block $[z]$ distinct from both that is between them.*

Proof. Suppose $x \otimes y$. As before, $x \oplus y$ is divisible by two (possibly with remainder): there is a $z \in |M|$ such that either $x \oplus y = z \oplus z$ or $x \oplus y = (z \oplus z)^*$. The element z is the "average" of x and y, and $x \otimes z$ and $z \otimes y$. □

The non-standard blocks are therefore ordered like the rationals: they form a countably infinite dense linear ordering without endpoints. One can show that any two such countably infinite orderings are isomorphic. It follows that for any two countable non-standard models M_1 and M_2 of true arithmetic, their reducts to the language containing $<$ and $=$ only are isomorphic. Indeed, an isomorphism h can be defined as follows: the standard parts of M_1 and M_2 are isomorphic to the standard model N and hence to each other. The blocks making up the non-standard part are themselves ordered like the rationals and therefore isomorphic; an isomorphism of the blocks can be extended to an isomorphism *within* the blocks by matching up arbitrary elements in each, and then taking the image of the successor of x in M_1 to be the successor of the image of x in M_2. Note that it does *not* follow that $\mathfrak{M}_1$ and $\mathfrak{M}_2$ are isomorphic in the full language of arithmetic (indeed, isomorphism is always relative to a language), as there are non-isomorphic ways to define addition and multiplication over $|M_1|$ and $|M_2|$. (This also follows from a famous theorem due to Vaught that the number of countable models of a complete theory cannot be 2.)

6.9 Computable Models of Arithmetic

The standard model N has two nice features. Its domain is the natural numbers $\mathbb{N}$, i.e., its elements are just the kinds of things we want to talk about using the language of arithmetic, and the standard numeral $\overline{n}$ actually picks out n. The other nice feature is that the interpretations of the non-logical symbols of $\mathcal{L}_A$ are all *computable*. The successor, addition, and multiplication functions which serve as $\prime^N$, $+^N$, and $\times^N$ are computable functions of numbers. (Computable by Turing machines, or definable by primitive recursion, say.) And the less-than relation on N, i.e., $<^N$, is decidable.

Non-standard models of arithmetical theories such as **Q** and **PA** must contain non-standard elements. Thus their domains typically include elements in addition to $\mathbb{N}$. However, any countable structure can be built on any countably infinite set, including $\mathbb{N}$. So there are also non-standard models with domain $\mathbb{N}$. In such models M, of course, at least some numbers cannot play the roles they usually play, since some k must be different from $\mathrm{Val}^{M}(\overline{n})$ for all $n \in \mathbb{N}$.

Definition 6.30. A structure M for $\mathcal{L}_A$ is *computable* iff $|M| = \mathbb{N}$ and $\prime^M$, $+^M$, $\times^M$ are computable functions and $<^M$ is a decidable relation.

Example 6.31. Recall the structure K from Example 6.18 Its domain was $|K| = \mathbb{N} \cup \{a\}$ and interpretations

$$\begin{aligned}
\mathrm{o}^K &= 0 \\
\prime^K(x) &= \begin{cases} x+1 & \text{if } x \in \mathbb{N} \\ a & \text{if } x = a \end{cases} \\
+^K(x,y) &= \begin{cases} x+y & \text{if } x, y \in \mathbb{N} \\ a & \text{otherwise} \end{cases} \\
\times^K(x,y) &= \begin{cases} xy & \text{if } x, y \in \mathbb{N} \\ a & \text{otherwise} \end{cases} \\
<^K &= \{\langle x,y\rangle : x,y \in \mathbb{N} \text{ and } x < y\} \cup \{\langle x,a\rangle : n \in |K|\}
\end{aligned}$$

But $|K|$ is countably infinite and so is equinumerous with $\mathbb{N}$. For instance, $g \colon \mathbb{N} \to |K|$ with $g(0) = a$ and $g(n) = n+1$ for $n > 0$ is a bijection. We can turn it into an isomorphism between a new model K' of $\mathbf{Q}$ and K. In K', we have to assign different functions and relations to the symbols of $\mathcal{L}_A$, since different elements of $\mathbb{N}$ play the roles of standard and non-standard numbers.

Specifically, 0 now plays the role of a, not of the smallest standard number. The smallest standard number is now 1. So we assign $\mathrm{o}^{K'} = 1$. The successor function is also different now: given a standard number, i.e., an $n > 0$, it still returns $n+1$. But 0 now plays the role of a, which is its own successor. So $\prime^{K'}(0) = 0$.

For addition and multiplication we likewise have

$$+^{K'}(x,y) = \begin{cases} x+y & \text{if } x, y > 0 \\ 0 & \text{otherwise} \end{cases}$$

$$\times^{K'}(x,y) = \begin{cases} xy & \text{if } x, y > 0 \\ 0 & \text{otherwise} \end{cases}$$

And we have $\langle x, y \rangle \in <^{K'}$ iff $x < y$ and $x > 0$ and $y > 0$, or if $y = 0$.

All of these functions are computable functions of natural numbers and $<^{K'}$ is a decidable relation on $\mathbb{N}$—but they are not the same functions as successor, addition, and multiplication on $\mathbb{N}$, and $<^{K'}$ is not the same relation as $<$ on $\mathbb{N}$.

This example shows that **Q** has computable non-standard models with domain $\mathbb{N}$. However, the following result shows that this is not true for models of **PA** (and thus also for models of **TA**).

Theorem 6.32 (Tennenbaum's Theorem). *N is the only computable model of* **PA**.

Summary

A **model of arithmetic** is a structure for the language $\mathcal{L}_A$ of arithmetic. There is one distinguished such model, the **standard model** N, with $|N| = \mathbb{N}$ and interpretations of o, $'$, $+$, $\times$, and $<$ given by 0, the successor, addition, and multiplication functions on $\mathbb{N}$, and the less-than relation. N is a model of the theories **Q** and **PA**.

More generally, a structure for $\mathcal{L}_A$ is called **standard** iff it is isomorphic to N. Two structures are isomorphic if there is an **isomorphism** between them, i.e., a bijective function which preserves the interpretations of constant symbols, function symbols, and predicate symbols. By the **isomorphism theorem**, isomorphic structures are **elementarily equivalent**, i.e., they make the same sentences true. In standard models, the domain is just the set of values of all the numerals $\overline{n}$.

Models of **Q** and **PA** that are not isomorphic to N are called **non-standard**. In non-standard models, the domain is not exhausted by the values of the numerals. An element $x \in |M|$ where $x \neq \mathrm{Val}^{M}(\overline{n})$ for all $n \in \mathbb{N}$ is called a **non-standard element** of M. If $M \vDash$ **Q**, non-standard elements must obey the axioms of **Q**, e.g., they have unique successors, they can be added and multiplied, and compared using $<$. The standard elements of M are all $<^{M}$ all the non-standard elements. Non-standard models exist because of the compactness theorem, and for **Q** they can relatively easily be given explicitly. Such models can be used to show that, e.g., **Q** is not strong enough to prove certain sentences, e.g., **Q** $\nvDash \forall x\, \forall y\, (x+y) = (y+x)$. This is done by defining a non-standard M in which non-standard elements don't obey the law of commutativity.

Non-standard models of PA cannot be so easily specified explicitly. By showing that **PA** proves certain sentences, we can investigate the structure of the non-standard part of a non-standard model of **PA**. If a non-standard model M of **PA** is countable, every non-standard element is part of a "block" of non-standard elements which are ordered like $\mathbb{Z}$ by $<^{M}$. These blocks themselves are arranged like $\mathbb{Q}$, i.e., there is no smallest or largest block, and there is always a block in between any two blocks.

Any countable model is isomorphic to one with domain $\mathbb{N}$. If the interpretations of $\prime$, $+$, $\times$, and $<$ in such a model are computable functions, we say it is a **computable model**. The standard model N is computable, since the successor, addition, and multiplication functions and the less-than relation on $\mathbb{N}$ are computable. It is possible to define computable non-standard models of **Q**, but N is the only computable model of **PA**. This is **Tannenbaum's Theorem**.

Problems

Problem 6.1. Prove Proposition 6.2.

Problem 6.2. Carry out the proof of (b) of Theorem 6.6 in detail. Make sure to note where each of the five properties characterizing isomorphisms of Definition 6.5 is used.

Problem 6.3. Show that for any structure M, if X is a definable subset of M, and h is an automorphism of M, then $X = \{h(x) : x \in X\}$ (i.e., X is fixed under h).

Problem 6.4. Show that the converse of Proposition 6.12 is false, i.e., give an example of a structure M with $|M| = \{\mathrm{Val}^M(\overline{n}) : n \in \mathbb{N}\}$ that is not isomorphic to N.

Problem 6.5. Recall that $\mathbf{Q}$ contains the axioms

$$\forall x\, \forall y\, (x' = y' \rightarrow x = y) \tag{Q_1}$$

$$\forall x\, \mathsf{0} \neq x' \tag{Q_2}$$

$$\forall x\, (x = \mathsf{0} \lor \exists y\, x = y') \tag{Q_3}$$

Give structures M_1, M_2, M_3 such that

1. $M_1 \vDash Q_1$, $M_1 \vDash Q_2$, $M_1 \nvDash Q_3$;
2. $M_2 \vDash Q_1$, $M_2 \nvDash Q_2$, $M_2 \vDash Q_3$; and
3. $M_3 \nvDash Q_1$, $M_3 \vDash Q_2$, $M_3 \vDash Q_3$;

Obviously, you just have to specify $\mathsf{0}^{M_i}$ and $\prime^{M_i}$ for each.

Problem 6.6. Prove that K from Example 6.18 satisifies the remaining axioms of $\mathbf{Q}$,

$$\forall x\, (x \times \mathsf{0}) = \mathsf{0} \tag{Q_6}$$

$$\forall x\, \forall y\, (x \times y') = ((x \times y) + x) \tag{Q_7}$$

$$\forall x\, \forall y\, (x < y \leftrightarrow \exists z\, (z' + x) = y) \tag{Q_8}$$

Find a sentence only involving $\prime$ true in N but false in K.

Problem 6.7. Expand L of Example 6.19 to include $\otimes$ and $\otimes$ that interpret $\times$ and $<$. Show that your structure satisifies the remaining axioms of **Q**,

$$\forall x\,(x \times 0) = 0 \tag{Q_6}$$

$$\forall x\,\forall y\,(x \times y') = ((x \times y) + x) \tag{Q_7}$$

$$\forall x\,\forall y\,(x < y \leftrightarrow \exists z\,(z' + x) = y) \tag{Q_8}$$

Problem 6.8. In L of Example 6.19, $a^* = a$ and $b^* = b$. Is there a model of **Q** in which $a^* = b$ and $b^* = a$?

Problem 6.9. Find sentences in $\mathcal{L}_A$ derivable in **PA** (and hence true in N) which guarantee the properties of $\mathbf{z}$, $*$, and $\otimes$ in Proposition 6.21

Problem 6.10. Show that in a non-standard model of **PA**, there is no largest block.

Problem 6.11. Write out a detailed proof of Proposition 6.29. Which sentence must **PA** derive in order to guarantee the existence of z? Why is $x \otimes z$ and $z \otimes y$, and why is $[x] \neq [z]$ and $[z] \neq [y]$?

Problem 6.12. Give a structure L' with $|L'| = \mathbb{N}$ isomorphic to L of Example 6.19.

CHAPTER 7

Second-Order Logic

7.1 Introduction

In first-order logic, we combine the non-logical symbols of a given language, i.e., its constant symbols, function symbols, and predicate symbols, with the logical symbols to express things about first-order structures. This is done using the notion of satisfaction, which relates a structure M, together with a variable assignment s, and a formula A: $M, s \vDash A$ holds iff what A expresses when its constant symbols, function symbols, and predicate symbols are interpreted as M says, and its free variables are interpreted as s says, is true. The interpretation of the identity predicate $=$ is built into the definition of $M, s \vDash A$, as is the interpretation of $\forall$ and $\exists$. The former is always interpreted as the identity relation on the domain $|M|$ of the structure, and the quantifiers are always interpreted as ranging over the entire domain. But, crucially, quantification is only allowed over elements of the domain, and so only object variables are allowed to follow a quantifier.

In second-order logic, both the language and the definition of satisfaction are extended to include free and bound function and predicate variables, and quantification over them. These variables are related to function symbols and predicate symbols the same way that object variables are related to constant symbols. They play the same role in the formation of terms and formulas of second-order logic, and quantification over them is handled in a similar way. In the *standard* semantics, the second-order quantifiers range over all possible objects of the right type (n-place functions from $|M|$ to $|M|$ for function variables, n-place relations for predicate variables). For instance, while $\forall v_0 (P^1_0(v_0) \lor \lnot P^1_0(v_0))$ is a formula in both first- and second-order logic, in the latter we can also consider $\forall V^1_0 \forall v_0 (V^1_0(v_0) \lor \lnot V^1_0(v_0))$ and $\exists V^1_0 \forall v_0 (V^1_0(v_0) \lor \lnot V^1_0(v_0))$. Since these contain no free varaibles, they are sentences of second-order logic. Here, V^1_0 is a second-order 1-place predicate variable. The allowable interpretations of V^1_0 are the same that we can assign to a 1-place predicate symbol like P^1_0, i.e., subsets of $|M|$. Quantification over them then amounts to saying that $\forall v_0 (V^1_0(v_0) \lor \lnot V^1_0(v_0))$ holds for all ways of assigning a subset of $|M|$ as the value of V^1_0, or for at least one. Since every set either contains or fails to contain a given object, both are true in any structure.

Since second-order logic can quantify over subsets of the domain as well as functions, it is to be expected that some amount, at least, of set theory can be carried out in second-order logic. By "carry out," we mean that it is possible to express set theoretic properties and statements in second-order logic, and is possible without any special, non-logical vocabulary for sets (e.g., the membership predicate symbol of set theory). For instance, we can define unions and intersections of sets and the subset relationship, but also compare the sizes of sets, and state results such as Cantor's Theorem.

7.2 Terms and Formulas

Like in first-order logic, expressions of second-order logic are built up from a basic vocabulary containing *variables, constant symbols, predicate symbols* and sometimes *function symbols.* From them, together with logical connectives, quantifiers, and punctuation symbols such as parentheses and commas, *terms* and *formulas* are formed. The difference is that in addition to variables for objects, second-order logic also contains variables for relations and functions, and allows quantification over them. So the logical symbols of second-order logic are those of first-order logic, plus:

1. A countably infinite set of second-order relation variables of every arity n: $V_0^n, V_1^n, V_2^n, \ldots$

2. A countably infinite set of second-order function variables: $u_0^n, u_1^n, u_2^n, \ldots$

Just as we use x, y, z as meta-variables for first-order variables v_i, we'll use X, Y, Z, etc., as metavariables for V_i^n and u, v, etc., as meta-variables for u_i^n.

The non-logical symbols of a second-order language are specified the same way a first-order language is: by listing its constant symbols, function symbols, and predicate symbols

In first-order logic, the identity predicate $=$ is usually included. In first-order logic, the non-logical symbols of a language $\mathcal{L}$ are crucial to allow us to express anything interesting. There are of course sentences that use no non-logical symbols, but with only $=$ it is hard to say anything interesting. In second-order logic, since we have an unlimited supply of relation and function variables, we can say anything we can say in a first-order language even without a special supply of non-logical symbols.

Definition 7.1 (Second-order Terms). The set of *second-order terms* of $\mathcal{L}$, $\mathrm{Trm}^2(\mathcal{L})$, is defined by adding to Definition B.4 the clause

1. If u is an n-place function variable and $t_1, \ldots, t_n$ are terms, then $u(t_1, \ldots, t_n)$ is a term.

So, a second-order term looks just like a first-order term, except that where a first-order term contains a function symbol f_i^n, a second-order term may contain a function variable u_i^n in its place.

Definition 7.2 (Second-order formula). The set of *second-order formulas* $\mathrm{Frm}^2(\mathcal{L})$ of the language $\mathcal{L}$ is defined by adding to Definition B.4 the clauses

1. If X is an n-place predicate variable and $t_1, \ldots, t_n$ are second-order terms of $\mathcal{L}$, then $X(t_1, \ldots, t_n)$ is an atomic formula.

2. If A is a formula and u is a function variable, then $\forall u\, A$ is a formula.

3. If A is a formula and X is a predicate variable, then $\forall X\, A$ is a formula.

4. If A is a formula and u is a function variable, then $\exists u\, A$ is a formula.

5. If A is a formula and X is a predicate variable, then $\exists X\, A$ is a formula.

7.3 Satisfaction

To define the satisfaction relation $M, s \vDash A$ for second-order formulas, we have to extend the definitions to cover second-order variables. The notion of a structure is the same for second-order logic as it is for first-order logic. There is only a diffence for variable assignments s: these now must not just provide values for the first-order variables, but also for the second-order variables.

Definition 7.3 (Variable Assignment). A *variable assignment* s for a structure M is a function which maps each

1. object variable v_i to an element of $|M|$, i.e., $s(v_i) \in |M|$

2. n-place relation variable V_i^n to an n-place relation on $|M|$, i.e., $s(V_i^n) \subseteq |M|^n$;

3. n-place function variable u_i^n to an n-place function from $|M|$ to $|M|$, i.e., $s(u_i^n)\colon |M|^n \to |M|$;

A structure assigns a value to each constant symbol and function symbol, and a second-order variable assigns objects and functions to each object and function variable. Together, they let us assign a value to every term.

Definition 7.4 (Value of a Term). If t is a term of the language $\mathcal{L}$, M is a structure for $\mathcal{L}$, and s is a variable assignment for M, the *value* $\mathrm{Val}_s^M(t)$ is defined as for first-order terms, plus the following clause:

$t \equiv u(t_1, \dots, t_n)$:

$$\mathrm{Val}_s^M(t) = s(u)(\mathrm{Val}_s^M(t_1), \dots, \mathrm{Val}_s^M(t_n)).$$

Definition 7.5 (x-Variant). If s is a variable assignment for a structure M, then any variable assignment s' for M which differs from s at most in what it assigns to x is called an *x-variant* of s. If s' is an x-variant of s we write $s \sim_x s'$. (Similarly for second-order variables X or u.)

Definition 7.6 (Satisfaction). For second-order formulas A, the definition of satisfaction is like Definition B.23 with the addition of:

1. $A \equiv X^n(t_1, \dots, t_n)$: $M, s \vDash A$ iff $\langle \mathrm{Val}_s^M(t_1), \dots, \mathrm{Val}_s^M(t_n)\rangle \in s(X^n)$.

2. $A \equiv \forall X\, B$: $M, s \vDash A$ iff for every X-variant s' of s, $M, s' \vDash B$.

3. $A \equiv \exists X\, B$: $M, s \vDash A$ iff there is an X-variant s' of s so that $M, s' \vDash B$.

4. $A \equiv \forall u\, B$: $M, s \vDash A$ iff for every u-variant s' of s, $M, s' \vDash B$.

5. $A \equiv \exists u\, B$: $M, s \vDash A$ iff there is an u-variant s' of s so that $M, s' \vDash B$.

Example 7.7. Consider the formula $\forall z\, (X(z) \leftrightarrow \neg Y(z))$. It contains no second-order quantifiers, but does contain the second-order variables X and Y (here understood to be one-place). The corresponding first-order sentence $\forall z\, (P(z) \leftrightarrow \neg R(z))$ says that whatever falls under the interpretation of P does not fall under the interpretation of R and vice versa. In a structure, the interpretation of a predicate symbol P is given by the interpretation P^M. But for second-order variables like X and Y, the interpretation is provided, not by the structure itself, but by a variable assignment. Since the second-order formula is not a sentence (in includes free variables X and Y), it is only satisfied relative to a structure M together with a variable assignment s.

$M, s \vDash \forall z\, (X z \leftrightarrow \neg Y z)$ whenever the elements of $s(X)$ are not elements of $s(Y)$, and vice versa, i.e., iff $s(Y) = |M| \setminus s(X)$. So for instance, take $|M| = \{1, 2, 3\}$. Since no predicate symbols, function symbols, or constant symbols are involved, the domain of M is all that is relevant. Now for $s_1(X) = \{1, 2\}$ and $s_1(Y) = \{3\}$, we have $M, s_1 \vDash \forall z\, (X(z) \leftrightarrow \neg Y(z))$.

By contrast, if we have $s_2(X) = \{1, 2\}$ and $s_2(Y) = \{2, 3\}$, $M, s_2 \nvDash \forall z\, (X(z) \leftrightarrow \neg Y(z))$. That's because there is a z-variant s_2' of s_2 with $s_2'(z) = 2$ where $M, s_2' \vDash X(z)$ (since $2 \in s_2'(X)$) but $M, s_2' \nvDash \neg Y(z)$ (since also $s_2'(z) \in s_2'(Y)$).

Example 7.8. $M, s \vDash \exists Y\, (\exists y\, Y(y) \land \forall z\, (X(z) \leftrightarrow \neg Y(z)))$ if there is an $s' \sim_Y s$ such that $M, s' \vDash (\exists y\, Y(y) \land \forall z\, (X(z) \leftrightarrow \neg Y(z)))$. And that is the case iff $s'(Y) \neq \emptyset$ (so that $M, s' \vDash \exists y\, Y(y)$) and, as in the previous example, $s'(Y) = |M| \setminus s'(X)$. In other words, $M, s \vDash \exists Y\, (\exists y\, Y(y) \land \forall z\, (X(z) \leftrightarrow \neg Y(z)))$ iff $|M| \setminus s(X)$ is non-empty, i.e.,

$s(X) \neq |M|$. So, the formula is satisfied, e.g., if $|M| = \{1, 2, 3\}$ and $s(X) = \{1, 2\}$, but not if $s(X) = \{1, 2, 3\} = |M|$.

Since the formula is not satisfied whenever $s(X) = |M|$, the sentence

$$\forall X\, \exists Y\, (\exists y\, Y(y) \land \forall z\, (X(z) \leftrightarrow \lnot Y(z)))$$

is never satisfied: For any structure M, the assignment $s(X) = |M|$ will make the sentence false. On the other hand, the sentence

$$\exists X\, \exists Y\, (\exists y\, Y(y) \land \forall z\, (X(z) \leftrightarrow \lnot Y(z)))$$

is satisfied relative to any assignment s, since we can always find an X-variant s' of s with $s'(X) \neq |M|$.

7.4 Semantic Notions

The central logical notions of *validity*, *entailment*, and *satisfiability* are defined the same way for second-order logic as they are for first-order logic, except that the underlying satisfaction relation is now that for second-order formulas. A second-order sentence, of course, is a formula in which all variables, including predicate and function variables, are bound.

Definition 7.9 (Validity). A sentence A is *valid*, $\models A$, iff $M \models A$ for every structure M.

Definition 7.10 (Entailment). A set of sentences Γ *entails* a sentence A, $\Gamma \models A$, iff for every structure M with $M \models \Gamma$, $M \models A$.

Definition 7.11 (Satisfiability). A set of sentences Γ is *satisfiable* if $M \models \Gamma$ for some structure M. If Γ is not satisfiable it is called *unsatisfiable*.

7.5 Expressive Power

Quantification over second-order variables is responsible for an immense increase in the expressive power of the language over that of first-order logic. Second-order existential quantification lets us say that functions or relations with certain properties exists. In first-order logic, the only way to do that is to specify a non-logical symbol (i.e., a function symbol or predicate symbol) for this purpose. Second-order universal quantification lets us say that all subsets of, relations on, or functions from the domain to the domain have a property. In first-order logic, we can only say that the subsets, relations, or functions assigned to one of the non-logical symbols of the language have a property. And when we say that subsets, relations, functions exist that have a property, or that all of them have it, we can use second-order quantification in specifying this property as well. This lets us define relations not definable in first-order logic, and express properties of the domain not expressible in first-order logic.

Definition 7.12. If M is a structure for a language $\mathcal{L}$, a relation $R \subseteq |M|^2$ is *definable* in $\mathcal{L}$ if there is some formula $A_R(x, y)$ with only the variables x and y free, such that $R(a, b)$ holds (i.e., $\langle a, b \rangle \in R$) iff $M, s \vDash A_R(x, y)$ for $s(x) = a$ and $s(y) = b$.

Example 7.13. In first-order logic we can define the identity relation $\mathrm{Id}_{|M|}$ (i.e., $\{\langle a, a \rangle : a \in |M|\}$) by the formula $x = y$. In second-order logic, we can define this relation *without* $=$. For if a and b are the same element of $|M|$, then they are elements of the same subsets of $|M|$ (since sets are determined by their elements). Conversely, if a and b are different, then they are not elements of the same subsets: e.g., $a \in \{a\}$ but $b \notin \{a\}$ if $a \neq b$. So "being elements of the same subsets of $|M|$" is a relation that holds of a and b iff $a = b$. It is a relation that can be expressed in second-order logic, since we can quantify over all subsets of $|M|$. Hence, the following formula defines $\mathrm{Id}_{|M|}$:

$$\forall X\,(X(x) \leftrightarrow X(y))$$

Example 7.14. If R is a two-place predicate symbol, R^M is a two-place relation on $|M|$. Perhaps somewhat confusingly, we'll use R as the predicate symbol for R and for the relation R^M itself. The *transitive closure* R^* of R is the relation that holds between a and b iff for some $c_1, \dots, c_k$, $R(a, c_1)$, $R(c_1, c_2)$, ..., $R(c_k, b)$ holds. This includes the case if $k = 0$, i.e., if $R(a, b)$ holds, so does $R^*(a, b)$. This means that $R \subseteq R^*$. In fact, R^* is the smallest relation that includes R and that is transitive. We can say in second-order logic that X is a transitive relation that includes R:

$$B_R(X) \equiv \forall x\, \forall y\, (R(x, y) \rightarrow X(x, y)) \land \\ \forall x\, \forall y\, \forall z\, ((X(x, y) \land X(y, z)) \rightarrow X(x, z)).$$

The first conjunct says that $R \subseteq X$ and the second that X is transitive.

To say that X is the smallest such relation is to say that it is itself included in every relation that includes R and is transitive. So we can define the transitive closure of R by the formula

$$R^*(X) \equiv B_R(X) \land \forall Y\, (B_R(Y) \rightarrow \forall x\, \forall y\, (X(x, y) \rightarrow Y(x, y))).$$

We have $M, s \vDash R^*(X)$ iff $s(X) = R^*$. The transitive closure of R cannot be expressed in first-order logic.

7.6 Describing Infinite and Countable Domains

A set M is (Dedekind) infinite iff there is an injective function $f\colon M \rightarrow M$ which is not surjective, i.e., with $\mathrm{dom}(f) \neq M$. In first-order logic, we can consider a one-place function symbol f and say that the function f^M assigned to it in a structure M is injective and $\mathrm{ran}(f) \neq |M|$:

$$\forall x\, \forall y\, (f(x) = f(y) \rightarrow x = y) \land \exists y\, \forall x\, y \neq f(x).$$

If M satisfies this sentence, $f^M : |M| \rightarrow |M|$ is injective, and so $|M|$ must be infinite. If $|M|$ is infinite, and hence such a function exists, we can let f^M be that function and M will satisfy the sentence. However, this requires that our language contains the non-logical symbol f we use for this purpose. In second-order logic, we can simply say that such a function *exists*. This no-longer requires f, and we obtain the sentence in pure second-order logic

$$\mathrm{Inf} \equiv \exists u\, (\forall x\, \forall y\, (u(x) = u(y) \rightarrow x = y) \land \exists y\, \forall x\, y \neq u(x)).$$

$M \vDash \mathrm{Inf}$ iff $|M|$ is infinite. We can then define $\mathrm{Fin} \equiv \lnot\mathrm{Inf}$; $M \vDash \mathrm{Fin}$ iff $|M|$ is finite. No single sentence of pure first-order logic can express that the domain is infinite although an infinite set of them can. There is no set of sentences of pure first-order logic that is satisfied in a structure iff its domain is finite.

Proposition 7.15. *$M \vDash \mathrm{Inf}$ iff $|M|$ is infinite.*

Proof. $M \vDash \mathrm{Inf}$ iff $M, s \vDash \forall x\, \forall y\, (u(x) = u(y) \rightarrow x = y) \land \exists y\, \forall x\, y \neq u(x)$ for some s. If it does, $s(u)$ is an injective function, and some $y \in |M|$ is not in the domain of $s(u)$. Conversely, if there is an injective $f\colon |M| \rightarrow |M|$ with $\mathrm{dom}(f) \neq |M|$, then $s(u) = f$ is such a variable assignment. □

A set M is countable if there is an enumeration

$$m_0, m_1, m_2, \ldots$$

of its elements (without repetitions but possibly finite). Such an enumeration exists iff there is an element $z \in M$ and a function $f\colon M \rightarrow M$ such that z, $f(z)$, $f(f(z))$, …, are all the elements of M. For if the enumeration exists, $z = m_0$ and $f(m_k) = m_{k+1}$ (or $f(m_k) = m_k$ if m_k is the last element of the enumeration) are the requisite element and function. On the other hand, if such a z and f exist, then z, $f(z)$, $f(f(z))$, …, is an enumeration of M, and M is countable. We can express the existence of z and f in second-order logic to produce a sentence true in a structure iff the structure is countable:

$$\mathrm{Count} \equiv \exists z\, \exists u\, \forall X\, ((X(z) \land \forall x\, (X(x) \rightarrow X(u(x)))) \rightarrow \forall x\, X(x))$$

Proposition 7.16. *$M \vDash$ Count iff $|M|$ is countable.*

Proof. Suppose $|M|$ is countable, and let m_0, m_1, ..., be an enumeration. By removing repetions we can guarantee that no m_k appears twice. Define $f(m_k) = m_{k+1}$ and let $s(z) = m_0$ and $s(u) = f$. We show that

$$M, s \vDash \forall X\, ((X(z) \land \forall x\, (X(x) \to X(u(x)))) \to \forall x\, X(x))$$

Suppose $s' \sim_X s$ is arbitrary, and let $M = s'(X)$. Suppose further that $M, s' \vDash (X(z) \land \forall x\, (X(x) \to X(u(x))))$. Then $s'(z) \in M$ and whenever $x \in M$, also $s'(u)(x) \in M$. In other words, since $s' \sim_X s$, $m_0 \in M$ and if $x \in M$ then $f(x) \in M$, so $m_0 \in M$, $m_1 = f(m_0) \in M$, $m_2 = f(f(m_0)) \in M$, etc. Thus, $M = |M|$, and so $M \vDash \forall x\, X(x)s'$. Since s' was an arbitrary X-variant of s, we are done: $M \vDash$ Count.

Now assume that $M \vDash$ Count, i.e.,

$$M, s \vDash \forall X\, ((X(z) \land \forall x\, (X(x) \to X(u(x)))) \to \forall x\, X(x))$$

for some s. Let $m = s(z)$ and $f = s(u)$ and consider $M = \{m, f(m), f(f(m)), \dots\}$. Let s' be the X-variant of s with $s(X) = M$. Then

$$M, s' \vDash (X(z) \land \forall x\, (X(x) \to X(u(x)))) \to \forall x\, X(x)$$

by assumption. Also, $M, s' \vDash X(z)$ since $s'(X) = M \ni m = s'(z)$, and also $M, s' \vDash \forall x\, (X(x) \to X(u(x)))$ since whenever $x \in M$ also $f(x) \in M$. So, since both antecedent and conditional are satisfied, the consequent must also be: $M, s' \vDash \forall x\, X(x)$. But that means that $M = |M|$, and so $|M|$ is countable since M is, by definition. □

7.7 Second-order Arithmetic

Recall that the theory **PA** of Peano arithmetic includes the eight axioms of **Q**,

$$\forall x\, x' \neq 0$$

$$\forall x\, \forall y\, (x' = y' \rightarrow x = y)$$
$$\forall x\, (x = 0 \lor \exists y\, x = y')$$
$$\forall x\, (x + 0) = x$$
$$\forall x\, \forall y\, (x + y') = (x + y)'$$
$$\forall x\, (x \times 0) = 0$$
$$\forall x\, \forall y\, (x \times y') = ((x \times y) + x)$$
$$\forall x\, \forall y\, (x < y \leftrightarrow \exists z\, (z' + x) = y)$$

plus all sentences of the form

$$(A(0) \land \forall x\, (A(x) \rightarrow A(x'))) \rightarrow \forall x\, A(x).$$

The latter is a "schema," i.e., a pattern that generates infinitely many sentences of the language of arithmetic, one for each formula $A(x)$. We call this schema the (first-order) *axiom schema of induction.* In *second-order* Peano arithmetic $\mathbf{PA}^2$, induction can be stated as a single sentence. $\mathbf{PA}^2$ consists of the first eight axioms above plus the (second-order) *induction axiom*:

$$\forall X\, (X(0) \land \forall x\, (X(x) \rightarrow X(x'))) \rightarrow \forall x\, X(x)).$$

It says that if a subset X of the domain contains 0^M and with any $x \in |M|$ also contains ${\prime}^M(x)$ (i.e., it is "closed under successor") it contains everything in the domain (i.e., $X = |M|$).

The induction axiom guarantees that any structure satisfying it contains only those elements of $|M|$ the axioms require to be there, i.e., the values of $\overline{n}$ for $n \in \mathbb{N}$. A model of $\mathbf{PA}^2$ contains no non-standard numbers.

Theorem 7.17. *If* $M \vDash \mathbf{PA}^2$ *then* $|M| = \{\mathrm{Val}^M(\overline{n}) : n \in \mathbb{N}\}$.

Proof. Let $N = \{\mathrm{Val}^M(\overline{n}) : n \in \mathbb{N}\}$, and suppose $M \vDash \mathbf{PA}^2$. Of course, for any $n \in \mathbb{N}$, $\mathrm{Val}^M(\overline{n}) \in |M|$, so $N \subseteq |M|$.

Now for inclusion in the other direction. Consider a variable assignment s with $s(X) = N$. By assumption,

$$M \vDash \forall X\, (X(0) \land \forall x\, (X(x) \rightarrow X(x'))) \rightarrow \forall x\, X(x), \text{ thus}$$

$$M, s \vDash (X(0) \land \forall x\,(X(x) \to X(x'))) \to \forall x\, X(x).$$

Consider the antecedent of this conditional. $\mathrm{Val}^M(0) \in N$, and so $M, s \vDash X(0)$. The second conjunct, $\forall x\,(X(x) \to X(x'))$ is also satisfied. For suppose $x \in N$. By definition of N, $x = \mathrm{Val}^M(\overline{n})$ for some n. That gives $\prime^M(x) = \mathrm{Val}^M(\overline{n+1}) \in N$. So, $\prime^M(x) \in N$.

We have that $M, s \vDash X(0) \land \forall x\,(X(x) \to X(x'))$. Consequently, $M, s \vDash \forall x\, X(x)$. But that means that for every $x \in |M|$ we have $x \in s(X) = N$. So, $|M| \subseteq N$. □

Corollary 7.18. *Any two models of* $\mathbf{PA}^2$ *are isomorphic.*

Proof. By Theorem 7.17, the domain of any model of $\mathbf{PA}^2$ is exhausted by $\mathrm{Val}^M(\overline{n})$. Any such model is also a model of $\mathbf{Q}$. By Proposition 6.13, any such model is standard, i.e., isomorphic to N. □

Above we defined $\mathbf{PA}^2$ as the theory that contains the first eight arithmetical axioms plus the second-order induction axiom. In fact, thanks to the expressive power of second-order logic, only the *first two* of the arithmetical axioms plus induction are needed for second-order Peano arithmetic.

Proposition 7.19. *Let* $\mathbf{PA}^{2\dagger}$ *be the second-order theory containing the first two arithmetical axioms (the successor axioms) and the second-order induction axiom. Then* $\leq$, $+$, *and* $\times$ *are definable in* $\mathbf{PA}^{2\dagger}$.

Proof. To show that $\leq$ is definable, we have to find a formula $A_{\leq}(x, y)$ such that $N \vDash A(\overline{n}, \overline{m})$ iff $n < m$. Consider the formula

$$B(x, Y) \equiv Y(x) \land \forall y\,(Y(y) \to Y(y'))$$

Clearly, $B(\overline{n}, Y)$ is satisfied by a set $Y \subseteq \mathbb{N}$ iff $\{m : n \leq m\} \subseteq Y$, so we can take $A_{\leq}(x, y) \equiv \forall Y\,(B(x, Y) \to Y(y))$. □

Corollary 7.20. $M \vDash \mathbf{PA}^2$ *iff* $M \vDash \mathbf{PA}^{2\dagger}$.

Proof. Immediate from Proposition 7.19. □

7.8 Second-order Logic is not Axiomatizable

Theorem 7.21. *Second-order logic is undecidable.*

Proof. A first-order sentence is valid in first-order logic iff it is valid in second-order logic, and first-order logic is undecidable. □

Theorem 7.22. *There is no sound and complete proof system for second-order logic.*

Proof. Let A be a sentence in the language of arihmetic. $N \vDash A$ iff $\mathbf{PA}^2 \vDash A$. Let P be the conjunction of the nine axioms of $\mathbf{PA}^2$. $\mathbf{PA}^2 \vDash A$ iff $\vDash P \to A$, i.e., $M \vDash P \to A$. Now consider the sentence $\forall z\, \forall u\, \forall u'\, \forall u''\, \forall L\, (P' \to A')$ resulting by replacing o by z, $'$ by the one-place function variable u, $+$ and $\times$ by the two-place function-variables u' and u'', respectively, and $<$ by the two-place relation variable L and universally quantifying. It is a valid sentence of pure second-order logic iff the original sentence was valid iff $\mathbf{PA}^2 \vDash A$ iff $N \vDash A$. Thus if there were a sound and complete proof system for second-order logic, we could use it to define a computable enumeration $f : \mathbb{N} \to \mathrm{Sent}(\mathcal{L}_A)$ of the sentences true in N. This function would be representable in $\mathbf{Q}$ by some first-order formula $B_f(x, y)$. Then the formula $\exists x\, B_f(x, y)$ would define the set of true first-order sentences of N, contradicting Tarski's Theorem. □

7.9 Second-order Logic is not Compact

Call a set of sentences Γ *finitely satisfiable* if every one of its finite subsets is satisfiable. First-order logic has the property that if a set of sentences Γ is finitely satisfiable, it is satisfiable. This property is called *compactness*. It has an equivalent version involving entailment: if $\Gamma \vDash A$, then already $\Gamma_0 \vDash A$ for some finite subset $\Gamma_0 \subseteq \Gamma$. In this version it is an immediate corollary of the completeness theorem: for if $\Gamma \vDash A$, by completeness $\Gamma \vdash A$. But a derivation can only make use of finitely many sentences of Γ.

Compactness is not true for second-order logic. There are sets of second-order sentences that are finitely satisfiable but not satisfiable, and that entail some A without a finite subset entailing A.

Theorem 7.23. *Second-order logic is not compact.*

Proof. Recall that

$$\text{Inf} \equiv \exists u\, (\forall x\, \forall y\, (u(x) = u(y) \rightarrow x = y) \land \exists y\, \forall x\, y \neq u(x))$$

is satisfied in a structure iff its domain is infinite. Let $A^{\geq n}$ be a sentence that asserts that the domain has at least n elements, e.g.,

$$A^{\geq n} \equiv \exists x_1 \ldots \exists x_n\, (x_1 \neq x_2 \land x_1 \neq x_3 \land \cdots \land x_{n-1} \neq x_n).$$

Consider the set of sentences

$$\Gamma = \{\neg\text{Inf}, A^{\geq 1}, A^{\geq 2}, A^{\geq 3}, \ldots\}.$$

It is finitely satisfiable, since for any finite subset $\Gamma_0 \subseteq \Gamma$ there is some k so that $A^{\geq k} \in \Gamma$ but no $A^{\geq n} \in \Gamma$ for $n > k$. If $|M|$ has k elements, $M \vDash \Gamma_0$. But, Γ is not satisfiable: if $M \vDash \neg\text{Inf}$, $|M|$ must be finite, say, of size k. Then $M \nvDash A^{\geq k+1}$. □

7.10 The Löwenheim-Skolem Theorem Fails for Second-order Logic

The (Downward) Löwenheim-Skolem Theorem states that every set of sentences with an infinite model has a countable model. It, too, is a consequence of the completeneness theorem: the proof of completeness generates a model for any consistent set of sentences, and that model is countable. There is also an Upward Löwenheim-Skolem Theorem, which guarantees that if a set of sentences has a countably infinite model it also has an uncountable model. Both theorems fail in second-order logic.

Theorem 7.24. *The Löwenheim-Skolem Theorem fails for second-order logic: There are sentences with infinite models but no countable models.*

Proof. Recall that

$$\text{Count} \equiv \exists z\, \exists u\, \forall X\, ((X(z) \land \forall x\, (X(x) \rightarrow X(u(x)))) \rightarrow \forall x\, X(x))$$

is true in a structure M iff $|M|$ is countable, so $\neg$Count is true in M iff $|M|$ is uncountable. There are such structures—take any uncountable set as the domain, e.g., $\wp(\mathbb{N})$ or $\mathbb{R}$. So $\neg$Count has infinite models but no countable models. □

Theorem 7.25. *There are sentences with countably infinite but no uncountable models.*

Proof. Count $\land$ Inf is true in $\mathbb{N}$ but not in any structure M with $|M|$ uncountable. □

7.11 Comparing Sets

Proposition 7.26. *The formula $\forall x\, (X(x) \rightarrow Y(x))$ defines the subset relation, i.e., $M, s \vDash \forall x\, (X(x) \rightarrow Y(x))$ iff $s(X) \subseteq S(y)$.*

Proposition 7.27. *The formula $\forall x\, (X(x) \leftrightarrow Y(x))$ defines the identity relation on sets, i.e., $M, s \vDash \forall x\, (X(x) \leftrightarrow Y(x))$ iff $s(X) = S(y)$.*

Proposition 7.28. *The formula* $\exists x\, X(x)$ *defines the property of being non-empty, i.e.,* $\mathfrak{M}, s \vDash \exists x\, X(x)$ *iff* $s(X) \neq \emptyset$.

A set X is no larger than a set Y, $X \preceq Y$, iff there is an injective function $f\colon X \to Y$. Since we can express that a function is injective, and also that its values for arguments in X are in Y, we can also define the relation of being no larger than on subsets of the domain.

Proposition 7.29. *The formula*

$$\exists u\, (\forall x\, (X(x) \to Y(u(x))) \land \forall x\, \forall y\, (u(x) = u(y) \to x = y))$$

defines the relation of being no larger than.

Two sets are the same size, or "equinumerous," $X \approx Y$, iff there is a bijective function $f\colon X \to Y$.

Proposition 7.30. *The formula*

$$\begin{aligned} &\exists u\, (\forall x\, (X(x) \to Y(u(x))) \land \\ &\qquad \forall x\, \forall y\, (u(x) = u(y) \to x = y) \land \\ &\qquad\qquad \forall y\, (Y(y) \to \exists x\, (X(x) \land y = u(x)))) \end{aligned}$$

defines the relation of being equinumerous with.

We will abbreviate these formulas, respectively, as $X \subseteq Y$, $X = Y$, $X \neq \emptyset$, $X \preceq Y$, and $X \approx Y$. (This may be slightly confusing, since we use the same notation when we speak informally about sets X and Y—but here the notation is an abbreviation for formulas in second-order logic involving one-place relation variables X and Y.)

Proposition 7.31. *The sentence* $\forall X\,\forall Y\,((X \preceq Y \land Y \preceq X) \to X \approx Y)$ *is valid.*

Proof. The sentence is satisfied in a structure M if, for any subsets $X \subseteq |M|$ and $Y \subseteq |M|$, if $X \preceq Y$ and $Y \preceq X$ then $X \approx Y$. But this holds for *any* sets X and Y—it is the Schröder-Bernstein Theorem. □

7.12 Cardinalities of Sets

Just as we can express that the domain is finite or infinite, countable or uncountable, we can define the property of a subset of $|M|$ being finite or infinite, countable or uncountable.

Proposition 7.32. *The formula* $\mathrm{Inf}(X) \equiv$

$$\exists u\,(\forall x\,\forall y\,(u(x) = u(y) \to x = y) \land \\ \exists y\,(X(y) \land \forall x\,(X(x) \to y \neq u(x)))$$

is satisfied with respect to a variable assignment s *iff* $s(X)$ *is infinite.*

Proposition 7.33. *The formula* $\mathrm{Count}(X) \equiv$

$$\exists z\,\exists u\,(X(z) \land \forall x\,(X(x) \to X(u(x))) \land \\ \forall Y\,((Y(z) \land \forall x\,(Y(x) \to Y(u(x)))) \to X = Y))$$

is satisfied with respect to a variable assignment s *iff* $s(X)$ *is countable*

We know from Cantor's Theorem that there are uncountable sets, and in fact, that there are infinitely many different levels of infinite sizes. Set theory develops an entire arithmetic of sizes of sets, and assigns infinite cardinal numbers to sets. The natural numbers serve as the cardinal numbers measuring the sizes of finite sets. The cardinality of countably infinite sets is the first infinite cardinality, called $\aleph_0$ ("aleph-nought" or "aleph-zero").

The next infinite size is $\aleph_1$. It is the smallest size a set can be without being countable (i.e., of size $\aleph_0$). We can define "X has size $\aleph_0$" as $\mathrm{Aleph}_0(X) \leftrightarrow \mathrm{Inf}(X) \land \mathrm{Count}(X)$. X has size $\aleph_1$ iff all its subsets are finite or have size $\aleph_0$, but is not itself of size $\aleph_0$. Hence we can express this by the formula $\mathrm{Aleph}_1(X) \equiv \forall Y\, (Y \subseteq X \rightarrow (\lnot \mathrm{Inf}(Y) \lor \mathrm{Aleph}_0(Y))) \land \lnot \mathrm{Aleph}_0(X)$. Being of size $\aleph_2$ is defined similarly, etc.

There is one size of special interest, the so-called cardinality of the continuum. It is the size of $\wp(\mathbb{N})$, or, equivalently, the size of $\mathbb{R}$. That a set is the size of the continuum can also be expressed in second-order logic, but requires a bit more work.

7.13 The Power of the Continuum

In second-order logic we can quantify over subsets of the domain, but not over sets of subsets of the domain. To do this directly, we would need *third-order* logic. For instance, if we wanted to state Cantor's Theorem that there is no injective function from the power set of a set to the set itself, we might try to formulate it as "for every set X, and every set P, if P is the power set of X, then not $P \preceq X$. And to say that P is the power set of X would require formalizing that the elements of P are all and only the subsets of X, so something like $\forall Y\, (P(Y) \leftrightarrow Y \subseteq X)$. The problem lies in $P(Y)$: that is not a formula of second-order logic, since only terms can be arguments to one-place relation variables like P.

We can, however, *simulate* quantification over sets of sets, if the domain is large enough. The idea is to make use of the fact that two-place relations R relates elements of the domain to elements of the domain. Given such an R, we can collect all the elements to which some x is R-related: $\{y \in |M| : R(x, y)\}$ is the set "coded by" x. Converseley, if $Z \subseteq \wp(|M|)$ is some collection of subsets of $|M|$, and there are at least as many elements of $|M|$ as there are sets in Z, then there is also a relation $R \subseteq |M|^2$ such that every $Y \in Z$ is coded by some x using R.

Definition 7.34. If $R \subseteq |M|^2$, then x *R-codes* $\{y \in |M| : R(x,y)\}$. Y R-codes $\wp(X)$ iff for every $Z \subseteq X$, some $x \in Y$ R-codes Y, and every $x \in Y$ R-codes some $Y \in Z$.

Proposition 7.35. *The formula*

$$\text{Codes}(x, R, Y) \equiv \forall y\,(Y(y) \leftrightarrow R(x,y))$$

expresses that $s(x)$ $s(R)$*-codes* $s(Y)$. *The formula*

$$\begin{aligned}&\text{Pow}(Y, R, X) \equiv \\ &\qquad \forall Z\,(Z \subseteq X \rightarrow \exists x\,(Y(x) \land \text{Codes}(x, R, Z))) \land \\ &\qquad\qquad \forall x\,(Y(x) \rightarrow \forall Z\,(\text{Codes}(x, R, Z) \rightarrow Z \subseteq X)\end{aligned}$$

expresses that $s(Y)$ $s(R)$*-codes the power set of* $s(X)$.

With this trick, we can express statements about the power set by quantifying over the codes of subsets rather than the subsets themselves. For instance, Cantor's Theorem can now be expressed by saying that there is no injective function from the domain of any relation that codes the power set of X to X itself.

Proposition 7.36. *The sentence*

$$\begin{aligned}&\forall X\,\forall R\,(\text{Pow}(R, X) \rightarrow \\ &\qquad \neg\exists u\,(\forall x\,\forall y\,(u(x) = u(y) \rightarrow x = y) \land \\ &\qquad\qquad \forall Y\,(\text{Codes}(x, R, Y) \rightarrow X(u(x)))))\end{aligned}$$

is valid.

The power set of a countably infinite set is uncountable, and so its cardinality is larger than that of any countably infinite set (which is $\aleph_0$). The size of $\wp(\mathbb{R})$ is called the "power of the continuum," since it is the same size as the points on the real number line, $\mathbb{R}$. If the domain is large enough to code the power set of a countably infinite set, we can express that a set is the size of the continuum by saying that it is equinumerous with any set Y that codes the power set of set X of size $\aleph_0$. (If the domain is not large enough, i.e., it contains no subset equinumerous with $\mathbb{R}$, then there can also be no relation that codes $\wp(X)$.)

Proposition 7.37. *If* $\mathbb{R} \preceq |M|$, *then the formula*

$$\text{Cont}(X) \equiv \forall X\, \forall Y\, \forall R\, ((\text{Aleph}_0(X) \land \text{Pow}(Y, R, X)) \rightarrow \lnot Y \preceq X)$$

expresses that $s(X) \approx \mathbb{R}$.

Proposition 7.38. $|M| \approx \mathbb{R}$ *iff*

$$\begin{aligned} M \vDash \exists X\, \exists Y\, \exists R\, (&\text{Aleph}_0(X) \land \text{Pow}(Y, R, X) \land \\ &\exists u\, (\forall x\, \forall y\, (u(x) = u(y) \rightarrow x = y) \land \\ &\forall y\, (Y(y) \rightarrow \exists x\, y = u(x)))). \end{aligned}$$

The Continuum Hypothesis is the statement that the size of the continuum is the first uncountable cardinality, i.e, that $\wp(\mathbb{N})$ has size $\aleph_1$.

Proposition 7.39. *The Continuum Hypothesis is true iff*

$$\text{CH} \equiv \forall X\, (\text{Aleph}_1(X) \leftrightarrow \text{Cont}(x))$$

is valid.

Note that it isn't true that $\lnot$CH is valid iff the Continuum Hypothesis is false. In a countable domain, there are no subsets of size $\aleph_1$ and also no subsets of the size of the continuum, so CH is always true in a countable domain. However, we can give a different sentence that is valid iff the Continuum Hypothesis is false:

Proposition 7.40. *The Continuum Hypothesis is false iff*

$$\mathrm{NCH} \equiv \forall X\,(\mathrm{Cont}(X) \rightarrow \exists Y\,(Y \subseteq X \wedge \neg\mathrm{Count}(X) \wedge \neg X \approx Y))$$

is valid.

Summary

Second-order logic is an extension of first-order logic by variables for relations and functions, which can be quantified. Structures for second-order logic are just like first-order structures and give the interpretations of all non-logical symbols of the language. Variable assignments, however, also assign relations and functions on the domain to the second-order variables. The satisfaction relation is defined for second-order formulas just like in the first-order case, but extended to deal with second-order variables and quantifiers.

Second-order quantifiers make second-order logic **more expressive** than first-order logic. For instance, the identity relation on the domain of a structure can be defined without $=$, by $\forall X\,(X(x) \leftrightarrow X(y))$. Second-order logic can express the **transitive closure** of a relation, which is not expressible in first-order logic. Second-order quantifiers can also express properties of the domain, that it is finite or infinite, countable or uncountable. This means that, e.g., there is a second-order sentence Inf such that $M \vDash \mathrm{Inf}$ iff $|M|$ is infinite. Importantly, these are **pure** second-order sentences, i.e., they contain no non-logical symbols. Because of the compactness and Löwenheim-Skolem theorems, there are no first-order sentences that have these properties. It also shows that the **compactness and Löwenheim-Skolem theorems fail for second-order logic**.

Second-order quantification also makes it possible to replace first-order schemas by single sentences. For instance, **second-order arithmetic $\mathbf{PA}^2$** is comprised of the axioms of **Q** plus the single **induction axiom**

$$\forall X\,((X(0) \land \forall x\,(X(x) \to X(x'))) \to \forall x\,X(x)).$$

In contrast to first-order **PA**, all second-order models of $\mathbf{PA}^2$ are isomorphic to the standard model. In other words, $\mathbf{PA}^2$ has **no non-standard models**.

Since second-order logic includes first-order logic, it is undecidable. First-order logic is at least axiomatizable, i.e., it has a sound and complete proof system. Second-order logic does not, it is **not axiomatizable**. Thus, the set of validities of second-order logic is highly non-computable. In fact, pure second-order logic can express set-theoretic claims like the **continuum hypothesis**, which are independent of set theory.

Problems

Problem 7.1. Show that $\forall X\,(X(x) \to X(y))$ (note: $\to$ not $\leftrightarrow$!) defines $\mathrm{Id}_{|M|}$.

Problem 7.2. The sentence $\mathrm{Inf} \land \mathrm{Count}$ is true in all and only countably infinite domains. Adjust the definition of Count so that it becomes a different sentence that directly expresses that the domain is countably infinite, and prove that it does.

Problem 7.3. Complete the proof of Proposition 7.19.

Problem 7.4. Give an example of a set Γ and a sentence A so that $\Gamma \vDash A$ but for every finite subset $\Gamma_0 \subseteq \Gamma$, $\Gamma_0 \nvDash A$.

CHAPTER 8

The Lambda Calculus

8.1 Overview

The lambda calculus was originally designed by Alonzo Church in the early 1930s as a basis for constructive logic, and *not* as a model of the computable functions. But it was soon shown to be equivalent to other definitions of computability, such as the Turing computable functions and the partial recursive functions. The fact that this initially came as a small surprise makes the characterization all the more interesting.

Lambda notation is a convenient way of referring to a function directly by a symbolic expression which defines it, instead of defining a name for it. Instead of saying "let f be the function defined by $f(x) = x + 3$," one can say, "let f be the function $\lambda x.\,(x + 3)$." In other words, $\lambda x.\,(x + 3)$ is just a *name* for the function that adds three to its argument. In this expression, x is a dummy variable, or a placeholder: the same function can just as well be denoted by $\lambda y.\,(y + 3)$. The notation works even with other parameters around. For example, suppose $g(x, y)$ is a function of two variables, and k is a natural number. Then $\lambda x.\,g(x, k)$ is the function which maps any x to $g(x, k)$.

This way of defining a function from a symbolic expression is known as *lambda abstraction.* The flip side of lambda abstraction is *application*: assuming one has a function f (say, defined on the natural numbers), one can *apply* it to any value, like 2. In conventional notation, of course, we write $f(2)$ for the result.

What happens when you combine lambda abstraction with application? Then the resulting expression can be simplified, by "plugging" the applicand in for the abstracted variable. For example,

$$(\lambda x.\,(x+3))(2)$$

can be simplified to $2+3$.

Up to this point, we have done nothing but introduce new notations for conventional notions. The lambda calculus, however, represents a more radical departure from the set-theoretic viewpoint. In this framework:

1. Everything denotes a function.

2. Functions can be defined using lambda abstraction.

3. Anything can be applied to anything else.

For example, if F is a term in the lambda calculus, $F(F)$ is always assumed to be meaningful. This liberal framework is known as the *untyped* lambda calculus, where "untyped" means "no restriction on what can be applied to what."

There is also a *typed* lambda calculus, which is an important variation on the untyped version. Although in many ways the typed lambda calculus is similar to the untyped one, it is much easier to reconcile with a classical set-theoretic framework, and has some very different properties.

Research on the lambda calculus has proved to be central in theoretical computer science, and in the design of programming languages. LISP, designed by John McCarthy in the 1950s, is an early example of a language that was influenced by these ideas.

8.2 The Syntax of the Lambda Calculus

One starts with a sequence of variables x, y, z, ... and some constant symbols a, b, c, ... The set of terms is defined inductively, as follows:

1. Each variable is a term.

2. Each constant is a term.

3. If M and N are terms, so is (MN).

4. If M is a term and x is a variable, then $(\lambda x.\, M)$ is a term.

The system without any constants at all is called the *pure* lambda calculus.

We will follow a few notational conventions:

Convention 8.1.

1. When parentheses are left out, application takes place from left to right. For example, if M, N, P, and Q are terms, then $MNPQ$ abbreviates $(((MN)P)Q)$.

2. Again, when parentheses are left out, lambda abstraction is to be given the widest scope possible. From example, $\lambda x.\, MNP$ is read $\lambda x.\, (MNP)$.

3. A lambda can be used to abstract multiple variables. For example, $\lambda xyz.\, M$ is short for $\lambda x.\, \lambda y.\, \lambda z.\, M$.

For example,

$$\lambda xy.\, xxyx\lambda z.\, xz$$

abbreviates

$$\lambda x.\, \lambda y.\, ((((xx)y)x)\lambda z.\, (xz)).$$

You should memorize these conventions. They will drive you crazy at first, but you will get used to them, and after a while they will drive you less crazy than having to deal with a morass of parentheses.

Two terms that differ only in the names of the bound variables are called α-equivalent; for example, $\lambda x.x$ and $\lambda y.y$. It will be convenient to think of these as being the "same" term; in other words, when we say that M and N are the same, we also mean "up to renamings of the bound variables." Variables that are in the scope of a λ are called "bound", while others are called "free." There are no free variables in the previous example; but in

$$(\lambda z.yz)x$$

y and x are free, and z is bound.

8.3 Reduction of Lambda Terms

What can one do with lambda terms? Simplify them. If M and N are any lambda terms and x is any variable, we can use $M[N/x]$ to denote the result of substituting N for x in M, after renaming any bound variables of M that would interfere with the free variables of N after the substitution. For example,

$$(\lambda w.xxw)[yyz/x] = \lambda w.(yyz)(yyz)w.$$

Alternative notations for substitution are $[N/x]M$, $M[N/x]$, and also $M[x/N]$. Beware!

Intuitively, $(\lambda x.M)N$ and $M[N/x]$ have the same meaning; the act of replacing the first term by the second is called *β-contraction.* $(\lambda x.M)N$ is called a *redex* and $M[N/x]$ its *contractum.* Generally, if it is possible to change a term P to P' by β-contraction of some subterm, we say that *P β-reduces to P' in one step*, and write $P \longrightarrow P'$. If from P we can obtain P' with some number of one-step reductions (possibly none), then *P β-reduces to P'*; in symbols, $P \twoheadrightarrow P'$. A term that cannot be β-reduced any further is called *β-irreducible*, or *β-normal.* We will say "reduces" instead of "β-reduces," etc., when the context is clear.

Let us consider some examples.

1. We have

$$\begin{aligned}(\lambda x.\, xxy)\lambda z.\, z &\to (\lambda z.\, z)(\lambda z.\, z)y\\ &\to (\lambda z.\, z)y\\ &\to y.\end{aligned}$$

2. "Simplifying" a term can make it more complex:

$$\begin{aligned}(\lambda x.\, xxy)(\lambda x.\, xxy) &\to (\lambda x.\, xxy)(\lambda x.\, xxy)y\\ &\to (\lambda x.\, xxy)(\lambda x.\, xxy)yy\\ &\to \dots\end{aligned}$$

3. It can also leave a term unchanged:

$$(\lambda x.\, xx)(\lambda x.\, xx) \to (\lambda x.\, xx)(\lambda x.\, xx).$$

4. Also, some terms can be reduced in more than one way; for example,

$$(\lambda x.\, (\lambda y.\, yx)z)v \to (\lambda y.\, yv)z$$

by contracting the outermost application; and

$$(\lambda x.\, (\lambda y.\, yx)z)v \to (\lambda x.\, zx)v$$

by contracting the innermost one. Note, in this case, however, that both terms further reduce to the same term, zv.

The final outcome in the last example is not a coincidence, but rather illustrates a deep and important property of the lambda calculus, known as the "Church-Rosser property."

8.4 The Church-Rosser Property

Theorem 8.2. *Let M, N_1, and N_2 be terms, such that $M \twoheadrightarrow N_1$ and $M \twoheadrightarrow N_2$. Then there is a term P such that $N_1 \twoheadrightarrow P$ and $N_2 \twoheadrightarrow P$.*

Corollary 8.3. *Suppose M can be reduced to normal form. Then this normal form is unique.*

Proof. If $M \twoheadrightarrow N_1$ and $M \twoheadrightarrow N_2$, by the previous theorem there is a term P such that N_1 and N_2 both reduce to P. If N_1 and N_2 are both in normal form, this can only happen if $N_1 \equiv P \equiv N_2$.□

Finally, we will say that two terms M and N are *β-equivalent*, or just *equivalent*, if they reduce to a common term; in other words, if there is some P such that $M \twoheadrightarrow P$ and $N \twoheadrightarrow P$. This is written $M \overset{\beta}{=} N$. Using Theorem 8.2, you can check that $\overset{\beta}{=}$ is an equivalence relation, with the additional property that for every M and N, if $M \twoheadrightarrow N$ or $N \twoheadrightarrow M$, then $M \overset{\beta}{=} N$. (In fact, one can show that $\overset{\beta}{=}$ is the *smallest* equivalence relation having this property.)

8.5 Currying

A λ-abstract $\lambda x. M$ represents a function of one argument, which is quite a limitation when we want to define function accepting multiple arguments. One way to do this would be by extending the λ-calculus to allow the formation of pairs, triples, etc., in which case, say, a three-place function $\lambda x. M$ would expect its argument to be a triple. However, it is more convenient to do this by *Currying*.

Let's consider an example. If we want to define a function that accepts two arguments and returns the first, we write $\lambda x. \lambda y. x$, which literally is a function that accepts an argument and returns a function that accepts another argument and returns the first argument while it drops the second. Let's see what happens when we apply it to two arguments:

$$(\lambda x. \lambda y. x)MN \xrightarrow{\beta} (\lambda y. M)N$$
$$\xrightarrow{\beta} M$$

In general, to write a function with parameters $x_1, \ldots, x_n$ defined by some term N, we can write $\lambda x_1. \lambda x_2. \ldots \lambda x_n. N$. If we apply n arguments to it we get:

$$(\lambda x_1. \lambda x_2. \ldots \lambda x_n. N)M_1 \ldots M_n \xrightarrow{\beta}$$
$$\xrightarrow{\beta} ((\lambda x_2. \ldots \lambda x_n. N)[M_1/x_1])M_2 \ldots M_n$$
$$\equiv (\lambda x_2. \ldots \lambda x_n. N[M_1/x_1])M_2 \ldots M_n$$
$$\vdots$$
$$\xrightarrow{\beta} P[M_1/x_1] \ldots [M_n/x_n]$$

The last line literally means substituting M_i for x_i in the body of the function definition, which is exactly what we want when applying multiple arguments to a function.

8.6 Lambda Definability

At first glance, the lambda calculus is just a very abstract calculus of expressions that represent functions and applications of them to others. Nothing in the syntax of the lambda calculus suggests that these are functions of particular kinds of objects, in particular, the syntax includes no mention of natural numbers. Its basic operations—application and lambda abstractions—are operations that apply to any function, not just functions on natural numbers.

Nevertheless, with some ingenuity, it is possible to define arithmetical functions, i.e., functions on the natural numbers, in the lambda calculus. To do this, we define, for each natural number $n \in \mathbb{N}$, a special λ-term $\overline{n}$, the *Church numeral* for n. (Church numerals are named for Alonzo Church.)

Definition 8.4. If $n \in \mathbb{N}$, the corresponding *Church numeral* $\overline{n}$ represents n:

$$\overline{n} \equiv \lambda f x. f^n(x)$$

Here, $f^n(x)$ stands for the result of applying f to x n times. For example, $\overline{0}$ is $\lambda f x. x$, and $\overline{3}$ is $\lambda f x. f(f(f\, x))$.

The Church numeral $\overline{n}$ is encoded as a lambda term which represents a function accepting two arguments f and x, and returns $f^n(x)$. Church numerals are evidently in normal form.

A represention of natural numbers in the lambda calculus is only useful, of course, if we can compute with them. Computing with Church numerals in the lambda calculus means applying a λ-term F to such a Church numeral, and reducing the combined term $F\,\overline{n}$ to a normal form. If it always reduces to a normal form, and the normal form is always a Church numeral $\overline{m}$, we can think of the output of the computation as being the number m. We can then think of F as defining a function $f\colon \mathbb{N} \to \mathbb{N}$, namely the function such that $f(n) = m$ iff $F\,\overline{n} \twoheadrightarrow \overline{m}$. Because of the Church-Rosser property, normal forms are unique if they exist. So if $F\,\overline{n} \twoheadrightarrow \overline{m}$, there can be no other term in normal form, in particular no other Church numeral, that $F\,\overline{n}$ reduces to.

Conversely, given a function $f\colon \mathbb{N} \to \mathbb{N}$, we can ask if there is a term F that defines f in this way. In that case we say that F *λ-defines* f, and that f is λ-definable. We can generalize this to many-place and partial functions.

Definition 8.5. Suppose $f \colon \mathbb{N}^k \to \mathbb{N}$. We say that a lambda term F *λ-defines* f if for all $n_0, \dots, n_{k-1}$,

$$F\,\overline{n_0}\,\overline{m_1} \dots \overline{n_{k-1}} \twoheadrightarrow \overline{f(n_0, n_1, \dots, n_{k-1})}$$

if $f(n_0, \dots, n_{k-1})$ is defined, and $F\,\overline{n_0}\,\overline{n_1} \dots \overline{n_{k-1}}$ has no normal form otherwise.

A very simple example are the constant functions. The term $C_k \equiv \lambda x.\,\overline{k}$ λ-defines the function $c_k \colon \mathbb{N} \to \mathbb{N}$ such that $c(n) = k$. For $C_k\,\overline{n} \equiv (\lambda x.\,\overline{k})\overline{n} \to \overline{k}$ for any n. The identity function is λ-defined by $\lambda x.\,x$. More complex functions are of course harder to define, and often require a lot of ingenuity. So it is perhaps surprising that every computable function is λ-definable. The converse is also true: if a function is λ-definable, it is computable.

8.7 λ-Definable Arithmetical Functions

Proposition 8.6. *The successor function* succ *is λ-definable.*

Proof. A term that λ-defines the successor function is

$$\mathrm{Succ} \equiv \lambda a.\,\lambda f x.\,f(a f x).$$

Succ is a function that accepts as argument a number a, and evaluates to another function, $\lambda f x.\,f(a f x)$. That function is not itself a Church numeral. However, if the argument a is a Church numeral, it reduces to one. Consider:

$$(\lambda a.\,\lambda f x.\,f(a f x))\,\overline{n} \to \lambda f x.\,f(\overline{n} f x).$$

The embedded term $\overline{n} f x$ is a redex, since $\overline{n}$ is $\lambda f x.\,f^n x$. So $\overline{n} f x \to f^n x$ and so, for the entire term we have

$$\mathrm{Succ}\,\overline{n} \twoheadrightarrow \lambda f x.\,f(f^n(x)),$$

i.e., $\overline{n+1}$. □

Proposition 8.7. *The addition function* add *is λ-definable.*

Proof. Addition is λ-defined by the terms

$$\text{Add} \equiv \lambda ab.\, \lambda f x.\, af(bfx)$$

or, alternatively,

$$\text{Add}' \equiv \lambda ab.\, a \,\text{Succ}\, b.$$

The first addition works as follows: Add first accept two numbers a and b. The result is a function that accepts f and x and returns $af(bfx)$. If a and b are Church numerals $\overline{n}$ and $\overline{m}$, this reduces to $f^{n+m}(x)$, which is identical to $f^n(f^m(x))$. Or, slowly:

$$\begin{aligned}(\lambda ab.\, \lambda f x.\, af(bfx))\overline{n}\,\overline{m} &\to \lambda f x.\, \overline{n}\, f(\overline{m}\, f x) \\ &\to \lambda f x.\, \overline{n}\, f(f^m x) \\ &\to \lambda f x.\, f^n(f^m x) \equiv \overline{n+m}.\end{aligned}$$

The second representation of addition Add′ works differently: Applied to two Church numerals $\overline{n}$ and $\overline{m}$,

$$\text{Add}'\overline{n}\,\overline{m} \to \overline{n}\,\text{Succ}\,\overline{m}.$$

But $\overline{n} f x$ always reduces to $f^n(x)$. So,

$$\overline{n}\,\text{Succ}\,\overline{m} \twoheadrightarrow \text{Succ}^n(\overline{m}).$$

And since Succ λ-defines the successor function, and the successor function applied n times to m gives $n+m$, this in turn reduces to $\overline{n+m}$. □

Proposition 8.8. *Multiplication is λ-definable by the term*

$$\text{Mult} \equiv \lambda ab.\, \lambda f x.\, a(bf)x$$

Proof. To see how this works, suppose we apply Mult to Church numerals $\overline{n}$ and $\overline{m}$: Mult $\overline{n}\,\overline{m}$ reduces to $\lambda f x.\,\overline{n}(\overline{m}\,f)x$. The term $\overline{m}f$ defines a function which applies f to its argument m times. Consequently, $\overline{n}(\overline{m}f)x$ applies the function "apply f m times" itself n times to x. In other words, we apply f to x, $n \cdot m$ times. But the resulting normal term is just the Church numeral $\overline{nm}$. □

We can actually simplify this term further by η-reduction:

$$\text{Mult} \equiv \lambda ab.\,\lambda f.\,a(bf).$$

The definition of exponentiation as a λ-term is surprisingly simple:

$$\text{Exp} \equiv \lambda be.\,eb.$$

The first argument b is the base and the second e is the exponent. Intuitively, ef is f^e by our encoding of numbers. If you find it hard to understand, we can still define exponentiation also by iterated multiplication:

$$\text{Exp}' \equiv \lambda be.\,e(\text{Mult}\,b)\overline{1}.$$

Predecessor and subtraction on Church numeral is not as simple as we might think: it requires encoding of pairs.

8.8 Pairs and Predecessor

Definition 8.9. The pair of M and N (written $\langle M, N\rangle$) is defined as follows:

$$\langle M, N\rangle \equiv \lambda f.\,fMN.$$

Intuitively it is a function that accepts a function, and applies that function to the two elements of the pair. Following this idea we have this constructor, which takes two terms and returns the pair containing them:

$$\text{Pair} \equiv \lambda mn.\,\lambda f.\,fmn$$

Given a pair, we also want to recover its elements. For this we need two access functions, which accept a pair as argument and return the first or second elements in it:

$$\mathrm{Fst} \equiv \lambda p.\, p(\lambda mn.\, m)$$
$$\mathrm{Snd} \equiv \lambda p.\, p(\lambda mn.\, n)$$

Now with pairs we can λ-define the predecessor function:

$$\mathrm{Pred} \equiv \lambda n.\, \mathrm{Fst}(n(\lambda p.\, \langle \mathrm{Snd}\, p, \mathrm{Succ}(\mathrm{Snd}\, p)\rangle)\langle \overline{0}, \overline{0}\rangle)$$

Remember that $\overline{n}\, f x$ reduces to $f^n(x)$; in this case f is a function that accepts a pair p and returns a new pair containing the second component of p and the successor of the second component; x is the pair $\langle 0, 0\rangle$. Thus, the result is $\langle 0, 0\rangle$ for $n = 0$, and $\langle \overline{n-1}, \overline{n}\rangle$ otherwise. Pred then returns the first component of the result.

Subtraction can be defined as Pred applied to a, b times:

$$\mathrm{Sub} \equiv \lambda ab.\, b\mathrm{Pred}\, a.$$

8.9 Truth Values and Relations

We can encode truth values in the pure lambda calculus as follows:

$$\mathrm{true} \equiv \lambda x.\, \lambda y.\, x$$
$$\mathrm{false} \equiv \lambda x.\, \lambda y.\, y$$

Truth values are represented as *selectors*, i.e., functions that accept two arguments and returning one of them. The truth value true selects its first argument, and false its second. For example, true MN always reduces to M, while false MN always reduces to N.

Definition 8.10. We call a relation $R \subseteq \mathbb{N}^n$ λ-definable if there is a term R such that

$$R\, \overline{n_1} \ldots \overline{n_k} \xrightarrow{\beta}\!\!\!\!\rightarrow \mathrm{true}$$

whenever $R(n_1, \ldots, n_k)$ and

$$R\,\overline{n_1}\ldots\overline{n_k} \xrightarrow{\beta}\!\!\!\!\rightarrow \text{false}$$

otherwise.

For instance, the relation IsZero = {0} which holds of 0 and 0 only, is λ-definable by

$$\text{IsZero} \equiv \lambda n.\, n(\lambda x.\,\text{false})\,\text{true}.$$

How does it work? Since Church numerals are defined as iterators (functions which apply their first argument n times to the second), we set the initial value to be true, and for every step of iteration, we return false regardless of the result of the last iteration. This step will be applied to the initial value n times, and the result will be true if and only if the step is not applied at all, i.e., when $n = 0$.

On the basis of this representation of truth values, we can further define some truth functions. Here are two, the representations of negation and conjunction:

$$\text{Not} \equiv \lambda x.\, x\,\text{false}\,\text{true}$$
$$\text{And} \equiv \lambda x.\, \lambda y.\, xy\,\text{false}$$

The function "Not" accepts one argument, and returns true if the argument is false, and false if the argument is true. The function "And" accepts two truth values as arguments, and should return true iff both arguments are true. Truth values are represented as selectors (described above), so when x is a truth value and is applied to two arguments, the result will be the first argument if x is true and the second argument otherwise. Now And takes its two arguments x and y, and in return passes y and false to its first argument x. Assuming x is a truth value, the result will evaluate to y if x is true, and to false if x is false, which is just what is desired.

Note that we assume here that only truth values are used as arguments to And. If it is passed other terms, the result (i.e., the normal form, if it exists) may well not be a truth value.

8.10 Primitive Recursive Functions are λ-Definable

Recall that the primitive recursive functions are those that can be defined from the basic functions zero, succ, and P_i^n by composition and primitive recursion.

Lemma 8.11. *The basic primitive recursive functions* zero, succ, *and projections P_i^n are λ-definable.*

Proof. They are λ-defined by the following terms:

$$\begin{aligned} \text{Zero} &\equiv \lambda a.\, \lambda f x.\, x \\ \text{Succ} &\equiv \lambda a.\, \lambda f x.\, f(a f x) \\ \text{Proj}_i^n &\equiv \lambda x_0 \ldots x_{n-1}.\, x_i \end{aligned}$$

□

Lemma 8.12. *Suppose the k-ary function f, and n-ary functions $g_0, \ldots, g_{k-1}$, are λ-definable by terms F, G_0, ..., G_k, and h is defined from them by composition. Then H is λ-definable*

Proof. h can be λ-defined by the term

$$H \equiv \lambda x_0 \ldots x_{n-1}.\, F\,(G_0 x_0 \ldots x_{n-1}) \ldots (G_{k-1} x_0 \ldots x_{n-1})$$

We leave verification of this fact as an exercise. □

Note that Lemma 8.12 did not require that f and $g_0, \ldots, g_{k-1}$ are primitive recursive; it is only required that they are total and λ-definable.

Lemma 8.13. *Suppose f is an n-ary function and g is an $n+2$-ary function, they are λ-definable by terms F and G, and the function h is defined from f and g by primitive recursion. Then h is also λ-definable.*

Proof. Recall that h is defined by

$$h(x_1,\ldots,x_n,0) = f(x_1,\ldots,x_n)$$
$$h(x_1,\ldots,x_n,y+1) = h(x_1,\ldots,x_n,y,h(x_1,\ldots,x_n,y)).$$

Informally speaking, the primitive recursive definition iterates the application of the function h y times and applies it to $f(x_1,\ldots,x_n)$. This is reminiscent of the definition of Church numerals, which is also defined as a iterator.

For simplicity, we give the definition and proof for a single additional argument x. The function h is λ-defined by:

$$H \equiv \lambda x.\, \lambda y.\, \mathrm{Snd}(yD\langle \overline{0}, Fx\rangle)$$

where

$$D \equiv \lambda p.\, \langle \mathrm{Succ}(\mathrm{Fst}\, p), (Gx(\mathrm{Fst}\, p)(\mathrm{Snd}\, p))\rangle$$

The iteration state we maintain is a pair, the first of which is the current y and the second is the corresponding value of h. For every step of iteration we create a pair of new values of y and h; after the iteration is done we return the second part of the pair and that's the final h value. We now prove this is indeed a representation of primitive recursion.

We want to prove that for any n and m, $H\,\overline{n}\,\overline{m} \twoheadrightarrow \overline{h(n,m)}$. To do this we first show that if $D_n \equiv D[\overline{n}/x]$, then $D_n^m\langle \overline{0}, F\,\overline{n}\rangle \twoheadrightarrow \langle \overline{m}, \overline{h(n,m)}\rangle$ We proceed by induction on m.

If $m = 0$, we want $D_n^0\langle \overline{0}, F\,\overline{n}\rangle \twoheadrightarrow \langle \overline{0}, \overline{h(n,0)}\rangle$. But $D_n^0\langle \overline{0}, F\,\overline{n}\rangle$ just is $\langle \overline{0}, F\,\overline{n}\rangle$. Since F λ-defines f, this reduces to $\langle \overline{0}, \overline{f(n)}\rangle$, and since $f(n) = h(n,0)$, this is $\langle \overline{0}, \overline{h(n,0)}\rangle$

Now suppose that $D_n^m\langle \overline{0}, F\,\overline{n}\rangle \twoheadrightarrow \langle \overline{m}, \overline{h(n,m)}\rangle$. We want to show that $D_n^{m+1}\langle \overline{0}, F\,\overline{n}\rangle \twoheadrightarrow \langle \overline{m+1}, \overline{h(n,m+1)}\rangle$.

$$D_n^{m+1}\langle \overline{0}, F\,\overline{n}\rangle \equiv D_n(D_n^m\langle \overline{0}, F\,\overline{n}\rangle)$$

$$
\begin{aligned}
&\twoheadrightarrow D_n \, \langle \overline{m}, \overline{h(n,m)} \rangle \qquad \text{(by IH)} \\
&\equiv (\lambda p.\, \langle \text{Succ}(\text{Fst}\, p), (G\, \overline{n}(\text{Fst}\, p)(\text{Snd}\, p)) \rangle) \langle \overline{m}, \overline{h(n,m)} \rangle \\
&\rightarrow \langle \text{Succ}(\text{Fst}\, \langle \overline{m}, \overline{h(n,m)} \rangle), \\
&\qquad (G\, \overline{n}(\text{Fst}\, \langle \overline{m}, \overline{h(n,m)} \rangle)(\text{Snd}\, \langle \overline{m}, \overline{h(n,m)} \rangle)) \rangle \\
&\twoheadrightarrow \langle \text{Succ}\, \overline{m}, (G\, \overline{n}\, \overline{m}\, \overline{h(n,m)}) \rangle \\
&\twoheadrightarrow \langle \overline{m+1}, \overline{g(n,m,h(n,m))} \rangle
\end{aligned}
$$

Since $g(n,m,h(n,m)) = h(n,m+1)$, we are done.

Finally, consider

$$
\begin{aligned}
H\, \overline{n}\, \overline{m} &\equiv \lambda x.\, \lambda y.\, \text{Snd}(y(\lambda p. \langle \text{Succ}(\text{Fst}\, p), (G\, x\, (\text{Fst}\, p)\, (\text{Snd}\, p)) \rangle) \langle \overline{0}, F x \rangle) \\
&\quad \overline{n}\, \overline{m} \\
&\twoheadrightarrow \text{Snd}(\overline{m}\, \underbrace{(\lambda p. \langle \text{Succ}(\text{Fst}\, p), (G\, \overline{n}\, (\text{Fst}\, p)(\text{Snd}\, p)) \rangle)}_{D_n} \langle \overline{0}, F \overline{n} \rangle) \\
&\equiv \text{Snd}(\overline{m}\, D_n\, \langle \overline{0}, F \overline{n} \rangle) \\
&\twoheadrightarrow \text{Snd}\, (D_n^m \langle \overline{0}, F \overline{n} \rangle) \\
&\twoheadrightarrow \text{Snd}\, \langle \overline{m}, \overline{h(n,m)} \rangle \\
&\twoheadrightarrow \overline{h(n,m)}.
\end{aligned}
$$

□

Proposition 8.14. *Every primitive recursive function is λ-definable.*

Proof. By Lemma 8.11, all basic functions are λ-definable, and by Lemma 8.12 and Lemma 8.13, the λ-definable functions are closed under composition and primitive recursion. □

8.11 Fixpoints

Suppose we wanted to define the factorial function by recursion as a term Fac with the following property:

$$\text{Fac} \equiv \lambda n.\, \text{IsZero}\, n\, \overline{1}(\text{Mult}\, n(\text{Fac}(\text{Pred}\, n)))$$

That is, the factorial of n is 1 if $n = 0$, and n times the factorial of $n - 1$ otherwise. Of course, we cannot define the term Fac this way since Fac itself occurs in the right-hand side. Such recursive definitions involving self-reference are not part of the lambda calculus. Defining a term, e.g., by

$$\text{Mult} \equiv \lambda ab.\, a(\text{Add}\, a)0$$

only involves previously defined terms in the right-hand side, such as Add. We can always remove Add by replacing it with its defining term. This would give the term Mult as a pure lambda term; if Add itself involved defined terms (as, e.g., Add′ does), we could continue this process and finally arrive at a pure lambda term.

However this is not true in the case of recursive definitions like the one of Fac above. If we replace the occurrence of Fac on the right-hand side with the definition of Fac itself, we get:

$$\begin{aligned}\text{Fac} \equiv \lambda n.\, &\text{IsZero}\, n\, \overline{1}\\ &(\text{Mult}\, n((\lambda n.\, \text{IsZero}\, n\, \overline{1}\, (\text{Mult}\, n\, (\text{Fac}(\text{Pred}\, n))))(\text{Pred}\, n)))\end{aligned}$$

and we still haven't gotten rid of Fac on the right-hand side. Clearly, if we repeat this process, the definition keeps growing longer and the process never results in a pure lambda term. Thus this way of defining factorial (or more generally recursive functions) is not feasible.

The recursive definition does tell us something, though: If f were a term representing the factorial function, then the term

$$\text{Fac}' \equiv \lambda g.\, \lambda n.\, \text{IsZero}\, n\, \overline{1}\, (\text{Mult}\, n\, (g(\text{Pred} n)))$$

applied to the term f, i.e., $\text{Fac}'\, f$, also represents the factorial function. That is, if we regard Fac′ as a function accepting a function and returning a function, the value of $\text{Fac}'\, f$ is just f, provided f is the factorial. A function f with the property that $\text{Fac}'\, f \overset{\beta}{=} f$ is called a *fixpoint* of Fac′. So, the factorial is a fixpoint of Fac′.

There are terms in the lambda calculus that compute the fixpoints of a given term, and these terms can then be used to turn a term like Fac′ into the definition of the factorial.

Definition 8.15. The *Y-combinator* is the term:

$$Y \equiv (\lambda ux.\, x(uux))(\lambda ux.\, x(uux)).$$

Theorem 8.16. *Y has the property that $Y\,g \twoheadrightarrow g(Y\,g)$ for any term g. Thus, $Y\,g$ is always a fixpoint of g.*

Proof. Let's abbreviate $(\lambda ux.\, x(uux))$ by U, so that $Y \equiv UU$. Then

$$\begin{aligned} Y\,g &\equiv (\lambda ux.\, x(uux))U\,g \\ &\twoheadrightarrow (\lambda x.\, x(UUx))g \\ &\twoheadrightarrow g(UUg) \equiv g(Y\,g). \end{aligned}$$

Since $g(Y\,g)$ and $Y\,g$ both reduce to $g(Y\,g)$, $g(Y\,g) \stackrel{\beta}{=} Y\,g$, so $Y\,g$ is a fixpoint of g. □

Of course, since $Y\,g$ is a redex, the reduction can continue indefinitely:

$$\begin{aligned} Y\,g &\twoheadrightarrow g(Y\,g) \\ &\twoheadrightarrow g(g(Y\,g)) \\ &\twoheadrightarrow g(g(g(Y\,g))) \\ &\ldots \end{aligned}$$

So we can think of $Y\,g$ as g applied to itself infinitely many times. If we apply g to it one additional time, we—so to speak—aren't doing anything extra; g applied to g applied infinitely many times to $Y\,g$ is still g applied to $Y\,g$ infinitely many times.

Note that the above sequence of β-reduction steps starting with $Y\,g$ is infinite. So if we apply $Y\,g$ to some term, i.e., consider $(Y\,g)N$, that term will also reduce to infinitely many different terms, namely $(g(Y\,g))N$, $(g(g(Y\,g)))N$, ... It is nevertheless possible that some *other* sequence of reduction steps does terminate in a normal form.

Take the factorial for instance. Define Fac as $Y\,\text{Fac}'$ (i.e., a fixpoint of Fac′). Then:

$$\begin{aligned}
\text{Fac}\,\overline{3} &\twoheadrightarrow Y\,\text{Fac}'\,\overline{3}\\
&\twoheadrightarrow \text{Fac}'(Y\,\text{Fac}')\,\overline{3}\\
&\equiv (\lambda x.\,\lambda n.\,\text{IsZero}\,n\,\overline{1}\,(\text{Mult}\,n\,(x(\text{Pred}\,n))))\,\text{Fac}\,\overline{3}\\
&\twoheadrightarrow \text{IsZero}\,\overline{3}\,\overline{1}\,(\text{Mult}\,\overline{3}\,(\text{Fac}(\text{Pred}\,\overline{3})))\\
&\twoheadrightarrow \text{Mult}\,\overline{3}\,(\text{Fac}\,\overline{2}).
\end{aligned}$$

Similarly,

$$\begin{aligned}
\text{Fac}\,\overline{2} &\twoheadrightarrow \text{Mult}\,\overline{2}\,(\text{Fac}\,\overline{1})\\
\text{Fac}\,\overline{1} &\twoheadrightarrow \text{Mult}\,\overline{1}\,(\text{Fac}\,\overline{0})
\end{aligned}$$

but

$$\begin{aligned}
\text{Fac}\,\overline{0} &\twoheadrightarrow \text{Fac}'(Y\,\text{Fac}')\,\overline{0}\\
&\equiv (\lambda x.\,\lambda n.\,\text{IsZero}\,n\,\overline{1}\,(\text{Mult}\,n\,(x(\text{Pred}\,n))))\,\text{Fac}\,\overline{0}\\
&\twoheadrightarrow \text{IsZero}\,\overline{0}\,\overline{1}\,(\text{Mult}\,\overline{0}\,(\text{Fac}(\text{Pred}\,\overline{0}))).\\
&\twoheadrightarrow \overline{1}.
\end{aligned}$$

So together

$$\text{Fac}\,\overline{3} \twoheadrightarrow \text{Mult}\,\overline{3}\,(\text{Mult}\,\overline{2}\,(\text{Mult}\,\overline{1}\,\overline{1})).$$

What goes for Fac′ goes for any recursive definition. Suppose we have a recursive equation

$$g\,x_1 \ldots x_n \stackrel{\beta}{=} N$$

where N may contain g and $x_1, \ldots, x_n$. Then there is always a term $G \equiv (Y\lambda g.\,\lambda x_1 \ldots x_n.\,N)$ such that

$$G\,x_1 \ldots x_n \stackrel{\beta}{=} N[G/g].$$

For by the fixpoint theorem,

$$\begin{aligned} G \equiv (Y \lambda g.\, \lambda x_1 \ldots x_n.\, N) &\twoheadrightarrow \lambda g.\, \lambda x_1 \ldots x_n.\, N(Y \lambda g.\, \lambda x_1 \ldots x_n.\, N) \\ &\equiv (\lambda g.\, \lambda x_1 \ldots x_n.\, N)\, G \end{aligned}$$

and consequently

$$\begin{aligned} G\, x_1 \ldots x_n &\twoheadrightarrow (\lambda g.\, \lambda x_1 \ldots x_n.\, N)\, G\, x_1 \ldots x_n \\ &\twoheadrightarrow (\lambda x_1 \ldots x_n.\, N[G/g])\, x_1 \ldots x_n \\ &\twoheadrightarrow N[G/g]. \end{aligned}$$

The Y combinator of Definition 8.15 is due to Alan Turing. Alonzo Church had proposed a different version which we'll call Y_C:

$$Y_C \equiv \lambda g.\, (\lambda x.\, g(xx))(\lambda x.\, g(xx)).$$

Church's combinator is a bit weaker than Turing's in that $Y g \overset{\beta}{=} g(Y g)$ but not $Y g \overset{\beta}{\twoheadrightarrow} g(Y g)$. Let V be the term $\lambda x.\, g(xx)$, so that $Y_C \equiv \lambda g.\, V V$. Then

$$\begin{aligned} V V &\equiv (\lambda x.\, g(xx))V \twoheadrightarrow g(V V) \text{ and thus} \\ Y_C g &\equiv (\lambda g.\, V V)g \twoheadrightarrow V V \twoheadrightarrow g(V V), \text{ but also} \\ g(Y_C g) &\equiv g((\lambda g.\, V V)g) \twoheadrightarrow g(V V). \end{aligned}$$

In other words, $Y_C g$ and $g(Y_C g)$ reduce to a common term $g(V V)$; so $Y_C g \overset{\beta}{=} g(Y_C g)$. This is often enough for applications.

8.12 Minimization

The general recursive functions are those that can be obtained from the basic functions zero, succ, P_i^n by composition, primitive recursion, and regular minimization. To show that all general recursive functions are λ-definable we have to show that any function defined by regular minimization from a λ-definable function is itself λ-definable.

Lemma 8.17. *If $f(x_1, \ldots, x_k, y)$ is regular and λ-definable, then g defined by*

$$g(x_1, \ldots, x_k) = \mu y\ f(x_1, \ldots, x_k, y) = 0$$

is also λ-definable.

Proof. Suppose the lambda term F λ-defines the regular function $f(\vec{x}, y)$. To λ-define h we use a search function and a fixpoint combinator:

$$\text{Search} \equiv \lambda g.\, \lambda f\, \vec{x}\, y.\, \text{IsZero}(f\, \vec{x}\, y)\, y\, (g\, \vec{x}(\text{Succ}\, y)$$
$$H \equiv \lambda \vec{x}.\, (Y\, \text{Search}) F\, \vec{x}\, \overline{0},$$

where Y is any fixpoint combinator. Informally speaking, Search is a self-referencing function: starting with y, test whether $f\, \vec{x}\, y$ is zero: if so, return y, otherwise call itself with $\text{Succ}\, y$. Thus $(Y\, \text{Search}) F \overline{n_1} \ldots \overline{n_k}\, \overline{0}$ returns the least m for which $f(n_1, \ldots, n_k, m) = 0$.

Specifically, observe that

$$(Y\, \text{Search}) F \overline{n_1} \ldots \overline{n_k}\, \overline{m} \twoheadrightarrow \overline{m}$$

if $f(n_1, \ldots, n_k, m) = 0$, or

$$\twoheadrightarrow (Y\, \text{Search}) F\, \overline{n_1} \ldots \overline{n_k}\, \overline{m+1}$$

otherwise. Since f is regular, $f(n_1, \ldots, n_k, y) = 0$ for some y, and so

$$(Y\, \text{Search}) F \overline{n_1} \ldots \overline{n_k}\, \overline{0} \twoheadrightarrow \overline{h(n_1, \ldots, n_k)}. \qquad \square$$

Proposition 8.18. *Every general recursive function is λ-definable.*

Proof. By Lemma 8.11, all basic functions are λ-definable, and by Lemma 8.12, Lemma 8.13, and Lemma 8.17, the λ-definable functions are closed under composition, primitive recursion, and regular minimization. □

8.13 Partial Recursive Functions are λ-Definable

Partial recursive functions are those obtained from the basic functions by composition, primitive recursion, and unbounded minimization. They differ from general recursive function in that the functions used in unbounded search are not required to be regular. Not requiring regularity means that functions defined by minimization may sometimes not be defined.

At first glance it might seem that the same methods used to show that the (total) general recursive functions are all λ-definable can be used to prove that all partial recursive functions are λ-definable. For instance, the composition of f with g is λ-defined by $\lambda x. F(Gx)$ if f and g are λ-defined by terms F and G, respectively. However, when the functions are partial, this is problematic. When $g(x)$ is undefined, meaning Gx has no normal form. In most cases this means that $F(Gx)$ has no normal forms either, which is what we want. But consider when F is $\lambda x. \lambda y. y$, in which case $F(Gx)$ does have a normal form $(\lambda y. y)$.

This problem is not insurmountable, and there are ways to λ-define all partial recursive functions in such a way that undefined values are represented by terms without a normal form. These ways are, however, somewhat more complicated and less intuitive than the approach we have taken for general recursive functions. We record the theorem here without proof:

Theorem 8.19. *All partial recursive functions are λ-definable.*

8.14 λ-Definable Functions are Recursive

Not only are all partial recursive functions λ-definable, the converse is true, too. That is, all λ-definable functions are partial recursive.

Theorem 8.20. *If a partial function f is λ-definable, it is partial recursive.*

Proof. We only sketch the proof. First, we arithmetize λ-terms, i.e., systematially assign Gödel numbers to λ-terms as using the usual power-of-primes coding of sequences. Then we define a partial recursive function normalize(t) operating on the Gödel number t of a lambda term as argument, and which returns the Gödel number of the normal form if it has one, or is undefined otherwise. Then define two partial recursive functions toChurch and fromChurch that maps natural numbers to and from the Gödel numbers of the corresponding Church numeral.

Using these recursive functions, we can define the function f as a partial recursive function. There is a lambda term F that λ-defines f. To compute $f(n_1, \dots, n_k)$, first obtain the Gödel numbers of the corresponding Church numerals using toChurch(n_i), append these to ${}^{\#}F^{\#}$ to obtain the Gödel number of the term $F\overline{n_1} \dots \overline{n_k}$. Now use normalize on this Gödel number. If $f(n_1, \dots, n_k)$ is defined, $F\overline{n_1} \dots \overline{n_k}$ has a normal form (which must be a Church numeral), and otherwise it has no normal form (and so

$$\text{normalize}({}^{\#}F\overline{n_1} \dots \overline{n_k}{}^{\#})$$

is undefined). Finally, use fromChurch on the Gödel number of the normalized term. □

Problems

Problem 8.1. The term

$$\text{Succ}' \equiv \lambda n. \lambda f x. n f(f x)$$

λ-defines the successor function. Explain why.

Problem 8.2. Multiplication can be λ-defined by the term

$$\text{Mult}' \equiv \lambda ab. a(\text{Add } a)\overline{0}.$$

Explain why this works.

Problem 8.3. Explain why the access functions Fst and Snd work.

Problem 8.4. Define the functions Or and Xor representing the truth functions of inclusive and exclusive disjunction using the encoding of truth values as λ-terms.

Problem 8.5. Complete the proof of Lemma 8.12 by showing that $H\overline{n_0} \dots \overline{n_{n-1}} \twoheadrightarrow \overline{h(n_0, \dots, n_{n-1})}$.

APPENDIX A

Derivations in Arithmetic Theories

When we showed that all general recursive functions are representable in **Q**, and in the proofs of the incompleteness theorems, we claimed that various things are provable in **Q** and **PA**. The proofs of these claims, however, just gave the arguments informally without exhibiting actual derivations in natural deduction. We provide some of these derivations in this capter.

For instance, in Lemma 4.15 we proved that, for all n and $m \in \mathbb{N}$, $\mathbf{Q} \vdash (\overline{n} + \overline{m}) = \overline{n+m}$. We did this by induction on m.

Proof of Lemma 4.15. Base case: $m = 0$. Then what has to be proved is that, for all n, $\mathbf{Q} \vdash \overline{n} + \overline{0} = \overline{n+0}$. Since $\overline{0}$ is just o and $\overline{n+0}$ is $\overline{n}$, this amounts to showing that $\mathbf{Q} \vdash (\overline{n} + \mathrm{o}) = \overline{n}$. The derivation

$$\dfrac{\forall x\,(x + \mathrm{o}) = x}{(\overline{n} + \mathrm{o}) = \overline{n}}\ \forall\text{Elim}$$

is a natural deduction derivation of $(\overline{n} + \mathrm{o}) = \overline{n}$ with one undischarged assumption, and that undischarged assumption is an axiom of $\mathbf{Q}$.

Inductive step: Suppose that, for any n, $\mathbf{Q} \vdash (\overline{n} + \overline{m}) = \overline{n+m}$ (say, by a derivation $\delta_{n,m}$). We have to show that also $\mathbf{Q} \vdash (\overline{n} + \overline{m+1}) = \overline{n+m+1}$. Note that $\overline{m+1} \equiv \overline{m}'$, and that $\overline{n+m+1} \equiv \overline{n+m}'$. So we are looking for a derivation of $(\overline{n} + \overline{m}') = \overline{n+m}'$ from the axioms of $\mathbf{Q}$. Our derivation may use the derivation $\delta_{n,m}$ which exists by inductive hypothesis.

$$\dfrac{\dfrac{\vdots\,\delta_{n,m}}{(\overline{n} + \overline{m}) = \overline{n+m}} \qquad \dfrac{\dfrac{\forall x\,\forall y\,(x + y') = (x + y)'}{\forall y\,(\overline{n} + y') = (\overline{n} + y)'}\,\forall\text{Elim}}{(\overline{n} + \overline{m}') = (\overline{n} + \overline{m})'}\,\forall\text{Elim}}{(\overline{n} + \overline{m}') = \overline{n+m}'}\,{=}\text{Elim}$$

In the last =Elim inference, we replace the subterm $\overline{n} + \overline{m}$ of the right side $(\overline{n} + \overline{m})'$ of the right premise by the term $\overline{n+m}$. □

In Lemma 4.22, we showed that $\mathbf{Q} \vdash \forall x\, \neg x < 0$. What does an actual derivation look like?

Proof of Lemma 4.22. To prove a universal claim like this, we use $\forall$Intro, which requires a derivation of $\neg a < 0$. Looking at axiom Q_8, this means proving $\neg\exists z(z' + a) = 0$. Specifically, if we had a proof of the latter, Q_8 would allow us to prove the former (recall that $A \leftrightarrow B$ is short for $(A \to B) \wedge (B \to A)$.

$$\dfrac{\neg\exists z\,(z' + a) = 0 \qquad \dfrac{\dfrac{\dfrac{\dfrac{\forall x\,\forall y\,(x < y \leftrightarrow \exists z\,(z' + x) = y)}{\forall y\,(a < y \leftrightarrow \exists z\,(z' + a) = y)}\,\forall\text{Elim}}{a < 0 \leftrightarrow \exists z\,(z' + a) = 0}\,\forall\text{Elim}}{a < 0 \to \exists z\,(z' + a) = 0}\,\wedge\text{Elim} \qquad [a < 0]^1}{\exists z\,(z' + a) = 0}\,\to\text{Elim}}{1\ \dfrac{\bot}{\neg a < 0}\,\neg\text{Intro}}\,\neg\text{Elim}$$

This is a derivation of $\neg a < 0$ from $\neg\exists z\,(z' + a) = 0$ (and Q_8); let's call it δ_1.

Now how do we prove $\neg\exists z\,(z' + a) = 0$ from the axioms of $\mathbf{Q}$? To prove a negated claim like this, we'd need a derivation of the form

$$
{\scriptstyle 2}\ \dfrac{\begin{matrix}[\exists z\,(z' + a) = 0]^2 \\ \vdots \\ \bot\end{matrix}}{\neg\exists z\,(z' + a) = 0}\ \neg\text{Intro}
$$

To get a contradiction from an existential claim, we introduce a constant b for the existentially quantified variable z and use $\exists$Elim:

$$
{\scriptstyle 2}\ \dfrac{{\scriptstyle 3}\ \dfrac{[\exists z\,(z' + a) = 0]^2 \qquad \begin{matrix}[(b' + a) = 0]^3 \\ \vdots\,\delta_2 \\ \bot\end{matrix}}{\bot}\ \exists\text{Elim}}{\neg\exists z\,(z' + a) = 0}\ \neg\text{Intro}
$$

Now the task is to fill in δ_2, i.e., prove $\bot$ from $(b' + a) = 0$ and the axioms of **Q**. Q_2 says that 0 can't be the successor of some number, so one way of doing that would be to show that $(b' + a)$ is equal to the successor of some number. Since that expression itself is a sum, the axioms for addition must come into play. If $a = 0$, Q_5 would tell us that $(b'+a) = b'$, i.e., $b'+a$ is the successor of some number, namely of b. On the other hand, if $a = c'$ for some c, then $(b' + a) = (b' + c')$ by $=$Elim, and $(b' + c') = (b' + c)'$ by Q_6. So again, $b' + a$ is the successor of a number—in this case, $b'+c$. So the strategy is to divide the task into these two cases. We also have to verify that **Q** proves that one of these cases holds, i.e., $\mathbf{Q} \vdash a = 0 \lor \exists y\,(a = y')$, but this follows directly from Q_3 by $\forall$Elim. Here are the two cases:

Case 1: Prove $\bot$ from $a = 0$ and $(b' + a) = 0$ (and axioms Q_2, Q_5):

$$
\dfrac{\dfrac{\forall x\,\neg 0 = x'}{\neg 0 = b'}\ \forall\text{Elim} \qquad \dfrac{\dfrac{\forall x\,(x + 0) = x}{(b' + 0) = b'}\ \forall\text{Elim} \qquad \dfrac{\dfrac{a = 0 \qquad (b' + a) = 0}{(b' + 0) = 0}\ =\text{Elim}}{0 = (b' + 0)}}{0 = b'}\ =\text{Elim}}{\bot}\ \neg\text{Elim}
$$

Call this derivation δ_3. (We've abbreviated the derivation of $0 = (b' + 0)$ from $(b' + 0) = 0$ by a double inference line.)

Case 2: Prove $\bot$ from $\exists y\, a = y'$ and $(b' + a) = 0$ (and axioms Q_2, Q_6). We first show how to derive $\bot$ from $a = c'$ and $(b'+a) = 0$.

$$\dfrac{\dfrac{\forall x\, \neg 0 = x'}{\neg 0 = (b' + c)'}\;\forall\text{Elim} \qquad \dfrac{\dfrac{a = c' \quad (b' + a) = 0}{(b' + c') = 0}\;{=}\text{Elim} \qquad \dfrac{\dfrac{\forall x\, \forall y\, (x + y') = (x + y)'}{\forall y\, (b' + y') = (b' + y)'}\;\forall\text{Elim}}{(b' + c') = (b' + c)'}\;\forall\text{Elim}}{0 = (b' + c)'}\;{=}\text{Elim}}{\bot}\;\neg\text{Elim}$$

Call this δ_4. We get the required derivation δ_5 by applying $\exists$Elim and discharging the assumption $a = c'$:

$$\dfrac{\exists y\, a = y' \qquad \begin{array}{c} [a = c']^6 \quad (b' + a) = 0 \\ \vdots\ \delta_4 \\ \bot \end{array}}{\bot}\;{}^{6}\,\exists\text{Elim}$$

Putting everything together, the full proof looks like this:

$$\dfrac{\dfrac{\dfrac{[\exists z\, (z' + a) = 0]^2 \qquad \left.\dfrac{\dfrac{\forall x\, (x = 0 \lor \exists y\, (a = y'))}{a = 0 \lor \exists y\, (a = y')}\;\forall\text{Elim} \qquad \begin{array}{c} [a = 0]^7 \\ [(b' + a) = 0]^3 \\ \vdots\ \delta_3 \\ \bot \end{array} \qquad \begin{array}{c} [\exists y\, a = y']^7 \\ [(b' + a) = 0]^3 \\ \vdots\ \delta_5 \\ \bot \end{array}}{\bot}\;{}^{7}\,\lor\text{Elim}\right\}\delta_2}{\bot}\;{}^{3}\,\exists\text{Elim}}{\neg\exists z\, (z' + a) = 0}\;{}^{2}\,\neg\text{Intro}}{\begin{array}{c} \vdots\ \delta_1 \\ \neg a < 0 \end{array}}$$

$$\dfrac{\neg a < 0}{\forall x\, \neg x < 0}\;\forall\text{Intro}$$

□

In the proof of Theorem 5.7, we defined $\mathsf{RProv}(y)$ as

$$\exists x\,(\mathsf{Prf}(x,y) \land \forall z\,(z < x \rightarrow \lnot\mathsf{Ref}(z,y))).$$

$\mathsf{Prf}(x,y)$ is the formula representing the proof relation of $\mathbf{T}$ (a consistent, axiomatizable extension of $\mathbf{Q}$) in $\mathbf{Q}$, and $\mathsf{Ref}(z,y)$ is the formula representing the refutation relation. That means that if n is the Gödel number of a proof of A, then $\mathbf{Q} \vdash \mathsf{Prf}(\overline{n}, \ulcorner A \urcorner)$, and otherwise $\mathbf{Q} \vdash \lnot\mathsf{Prf}(\overline{n}, \ulcorner A \urcorner)$. Similarly, if n is the Gödel number of a proof of $\lnot A$, then $\mathbf{Q} \vdash \mathsf{Ref}(\overline{n}, \ulcorner A \urcorner)$, and otherwise $\mathbf{Q} \vdash \lnot\mathsf{Ref}(\overline{n}, \ulcorner A \urcorner)$. We use the Diagonal Lemma to find a sentence R such that $\mathbf{Q} \vdash R \leftrightarrow \lnot\mathsf{RProv}(\ulcorner R \urcorner)$. Rosser's Theorem states that $\mathbf{T} \nvdash R$ and $\mathbf{T} \nvdash \lnot R$. Both claims were proved indirectly: we show that if $\mathbf{T} \vdash R$, $\mathbf{T}$ is inconsistent, i.e., $\mathbf{T} \vdash \bot$, and the same if $\mathbf{T} \vdash \lnot R$.

Proof of Theorem 5.7. First we prove something things about $<$. By Lemma 4.23, we know that $\mathbf{Q} \vdash \forall x\,(x < \overline{n+1} \rightarrow (x = 0 \lor \cdots \lor x = \overline{n}))$ for every n. So of course also (if $n > 1$), $\mathbf{Q} \vdash \forall x\,(x < \overline{n} \rightarrow (x = 0 \lor \cdots \lor x = \overline{n-1}))$. We can use this to derive $a = 0 \lor \cdots \lor a = \overline{n-1}$ from $a < \overline{n}$:

$$\dfrac{a < \overline{n} \qquad \dfrac{\begin{array}{c}\vdots\\ \forall x\,(x < \overline{n} \rightarrow (x = \overline{0} \lor \cdots \lor x = \overline{n-1}))\end{array}}{a < \overline{n} \rightarrow (a = \overline{0} \lor \cdots \lor a = \overline{n-1})}\ \forall\text{Elim}}{a = \overline{0} \lor \cdots \lor a = \overline{n-1}}\ \rightarrow\text{Elim}$$

Let's call this derivation λ_1.

Now, to show that $\mathbf{T} \nvdash R$, we assume that $\mathbf{T} \vdash R$ (with a derivation δ) and show that $\mathbf{T}$ then would be inconsistent. Let n be the Gödel number of δ. Since Prf represents the proof relation in $\mathbf{Q}$, there is a derivation δ_1 of $\mathsf{Prf}(\overline{n}, \ulcorner R \urcorner)$. Furthermore, no $k < n$ is the Gödel number of a refutation of R since $\mathbf{T}$ is assumed to be consistent, so for each $k < n$, $\mathbf{Q} \vdash \lnot\mathsf{Ref}(\overline{k}, \ulcorner R \urcorner)$; let ρ_k be the corresponding derivation. We get a derivation of $\mathsf{RProv}(\ulcorner R \urcorner)$:

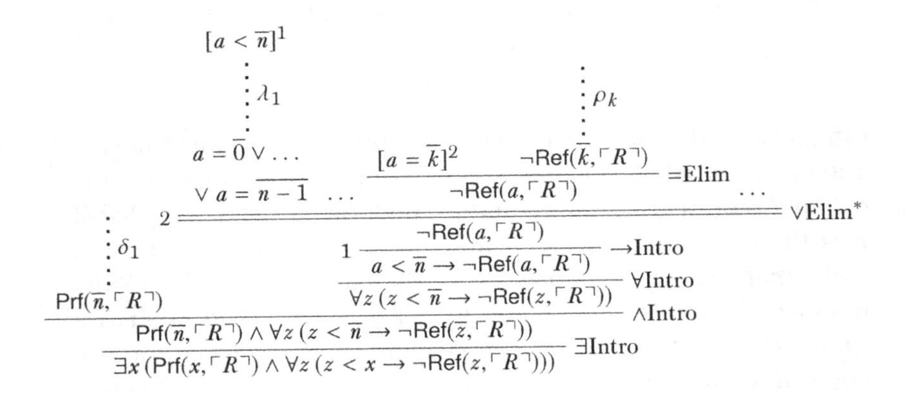

(We abbreviate multiple applications of $\lor$Elim by $\lor$Elim* above.) We've shown that if $\mathbf{T} \vdash R$ there would be a derivation of $\mathsf{RProv}(\ulcorner R \urcorner)$. Then, since $\mathbf{T} \vdash R \leftrightarrow \lnot\mathsf{RProv}(\ulcorner R \urcorner)$, also $\mathbf{T} \vdash \mathsf{RProv}(\ulcorner R \urcorner) \rightarrow \lnot R$, we'd have $\mathbf{T} \vdash \lnot R$ and $\mathbf{T}$ would be inconsistent.

Now let's show that $\mathbf{T} \nvdash \lnot R$. Again, suppose it did. Then there is a derivation ρ of $\lnot R$ with Gödel number m—a refutation of R—and so $\mathbf{Q} \vdash \mathsf{Ref}(\overline{m}, \ulcorner R \urcorner)$ by a derivation ρ_1. Since we assume $\mathbf{T}$ is consistent, $\mathbf{T} \nvdash R$. So for all k, k is not a Gödel number of a derivation of R, and hence $\mathbf{Q} \vdash \lnot\mathsf{Prf}(\overline{k}, \ulcorner R \urcorner)$ by a derivation π_k. So we have:

$$
\dfrac{\dfrac{\dfrac{\dfrac{\dfrac{\begin{matrix}\vdots\,\lambda_2\\ a=\overline{0}\vee\cdots\vee\\ a=\overline{m}\vee\overline{m}<a\end{matrix}\quad\ldots\quad\dfrac{\begin{matrix}[\mathsf{Prf}(a,\ulcorner R\urcorner)]^1\\ [a=\overline{k}]^2\\ \vdots\,\pi_k'\\ \bot\end{matrix}}{\overline{m}<a}\bot_I\quad\ldots\quad[\overline{m}<a]^2}{\overline{m}<a}\,{\scriptstyle 2}\ \vee\text{Elim}^* \qquad \begin{matrix}\vdots\,\rho_1\\ \mathsf{Ref}(\overline{m},\ulcorner R\urcorner)\end{matrix}}{\overline{m}<a\wedge\mathsf{Ref}(\overline{m},\ulcorner R\urcorner)}\wedge\text{Intro}}{\exists z(z<a\wedge\mathsf{Ref}(z,\ulcorner R\urcorner))}\exists\text{Intro}}{\mathsf{Prf}(a,\ulcorner R\urcorner)\to\exists z(z<a\wedge\mathsf{Ref}(z,\ulcorner R\urcorner))}\,{\scriptstyle 1}\to\text{Intro}}{\forall x(\mathsf{Prf}(x,\ulcorner R\urcorner)\to\exists z(z<x\wedge\mathsf{Ref}(z,\ulcorner R\urcorner)))}\forall\text{Intro}
$$

$$
\vdots
$$

$$
\neg\exists x(\mathsf{Prf}(x,\ulcorner R\urcorner)\wedge\forall z(z<x\to\neg\mathsf{Ref}(z,\ulcorner R\urcorner)))
$$

where π_k' is the derivation

$$
\dfrac{\begin{matrix}\vdots\,\pi_k\\ \neg\mathsf{Prf}(\overline{k},\ulcorner R\urcorner)\end{matrix}\qquad \dfrac{a=\overline{k}\qquad \mathsf{Prf}(a,\ulcorner R\urcorner)}{\mathsf{Prf}(\overline{k},\ulcorner R\urcorner)}=\text{Elim}}{\bot}\neg\text{Elim}
$$

and λ_2 is

$$
\dfrac{\begin{matrix}\vdots\,\lambda_3\\ (a<\overline{m}\vee\\ a=\overline{m})\vee\\ \overline{m}<a\end{matrix}\qquad \begin{matrix}[a<\overline{m}]^3\\ \vdots\,\lambda_1\\ a=\overline{0}\vee\cdots\vee\\ a=\overline{m-1}\\ \hline\hline a=\overline{0}\vee\cdots\vee\\ a=\overline{m}\vee\overline{m}<a\end{matrix}\qquad \begin{matrix}[a=\overline{m}]^3\\ \hline\hline a=\overline{0}\vee\cdots\vee\\ a=\overline{m}\vee\overline{m}<a\end{matrix}\qquad \begin{matrix}[\overline{m}<a]^3\\ \hline\hline a=\overline{0}\vee\cdots\vee\\ a=\overline{m}\vee\overline{m}<a\end{matrix}\ \vee\text{Intro}^*}{a=\overline{0}\vee\cdots\vee a=\overline{m}\vee\overline{m}<a}\,{\scriptstyle 3}\ \vee\text{Elim}^2
$$

(The derivation λ_3 exists by Lemma 4.24. We abbreviate repeated use of $\vee$Intro by $\vee$Intro* and the double use of $\vee$Elim to derive $a = \overline{0} \vee \cdots \vee a = \overline{m} \vee \overline{m} < a$ from $(a < \overline{m} \vee a = \overline{m}) \vee \overline{m} < a$ as $\vee$Elim^2.) □

APPENDIX B

First-order Logic

B.1 First-Order Languages

Expressions of first-order logic are built up from a basic vocabulary containing *variables*, *constant symbols*, *predicate symbols* and sometimes *function symbols*. From them, together with logical connectives, quantifiers, and punctuation symbols such as parentheses and commas, *terms* and *formulas* are formed.

Informally, predicate symbols are names for properties and relations, constant symbols are names for individual objects, and function symbols are names for mappings. These, except for the identity predicate =, are the *non-logical symbols* and together make up a language. Any first-order language $\mathscr{L}$ is determined by its non-logical symbols. In the most general case, $\mathscr{L}$ contains infinitely many symbols of each kind.

In the general case, we make use of the following symbols in first-order logic:

1. Logical symbols

    a) Logical connectives: $\neg$ (negation), $\wedge$ (conjunction), $\vee$ (disjunction), $\rightarrow$ (conditional), $\forall$ (universal quantifier), $\exists$ (existential quantifier).

b) The propositional constant for falsity $\bot$.

c) The two-place identity predicate $=$.

d) A countably infinite set of variables: v_0, v_1, v_2, ...

2. Non-logical symbols, making up the *standard language* of first-order logic

a) A countably infinite set of n-place predicate symbols for each $n > 0$: A^n_0, A^n_1, A^n_2, ...

b) A countably infinite set of constant symbols: c_0, c_1, c_2, ...

c) A countably infinite set of n-place function symbols for each $n > 0$: f^n_0, f^n_1, f^n_2, ...

3. Punctuation marks: (,), and the comma.

Most of our definitions and results will be formulated for the full standard language of first-order logic. However, depending on the application, we may also restrict the language to only a few predicate symbols, constant symbols, and function symbols.

Example B.1. The language $\mathcal{L}_A$ of arithmetic contains a single two-place predicate symbol $<$, a single constant symbol 0, one one-place function symbol $\prime$, and two two-place function symbols $+$ and $\times$.

Example B.2. The language of set theory $\mathcal{L}_Z$ contains only the single two-place predicate symbol $\in$.

Example B.3. The language of orders $\mathcal{L}_\leq$ contains only the two-place predicate symbol $\leq$.

Again, these are conventions: officially, these are just aliases, e.g., $<$, $\in$, and $\leq$ are aliases for A^2_0, 0 for c_0, $\prime$ for f^1_0, $+$ for f^2_0, $\times$ for f^2_1.

In addition to the primitive connectives and quantifiers introduced above, we also use the following *defined* symbols: $\leftrightarrow$ (biconditional), truth $\top$

A defined symbol is not officially part of the language, but is introduced as an informal abbreviation: it allows us to abbreviate formulas which would, if we only used primitive symbols, get quite long. This is obviously an advantage. The bigger advantage, however, is that proofs become shorter. If a symbol is primitive, it has to be treated separately in proofs. The more primitive symbols, therefore, the longer our proofs.

We might treat all the propositional operators and both quantifiers as primitive symbols of the language. We might instead choose a smaller stock of primitive symbols and treat the other logical operators as defined. "Truth functionally complete" sets of Boolean operators include $\{\neg, \lor\}$, $\{\neg, \land\}$, and $\{\neg, \rightarrow\}$—these can be combined with either quantifier for an expressively complete first-order language.

You may be familiar with two other logical operators: the Sheffer stroke $|$ (named after Henry Sheffer), and Peirce's arrow $\downarrow$, also known as Quine's dagger. When given their usual readings of "nand" and "nor" (respectively), these operators are truth functionally complete by themselves.

B.2 Terms and Formulas

Once a first-order language $\mathcal{L}$ is given, we can define expressions built up from the basic vocabulary of $\mathcal{L}$. These include in particular *terms* and *formulas*.

Definition B.4 (Terms). The set of *terms* $\text{Trm}(\mathcal{L})$ of $\mathcal{L}$ is defined inductively by:

1. Every variable is a term.

2. Every constant symbol of $\mathcal{L}$ is a term.

3. If f is an n-place function symbol and $t_1, \dots, t_n$ are terms, then $f(t_1, \dots, t_n)$ is a term.

4. Nothing else is a term.

A term containing no variables is a *closed term.*

The constant symbols appear in our specification of the language and the terms as a separate category of symbols, but they could instead have been included as zero-place function symbols. We could then do without the second clause in the definition of terms. We just have to understand $f(t_1, \dots, t_n)$ as just f by itself if $n = 0$.

Definition B.5 (Formula). The set of *formulas* $\mathrm{Frm}(\mathcal{L})$ of the language $\mathcal{L}$ is defined inductively as follows:

1. $\bot$ is an atomic formula.

2. If R is an n-place predicate symbol of $\mathcal{L}$ and $t_1, \dots, t_n$ are terms of $\mathcal{L}$, then $R(t_1, \dots, t_n)$ is an atomic formula.

3. If t_1 and t_2 are terms of $\mathcal{L}$, then $=(t_1, t_2)$ is an atomic formula.

4. If A is a formula, then $\neg A$ is formula.

5. If A and B are formulas, then $(A \land B)$ is a formula.

6. If A and B are formulas, then $(A \lor B)$ is a formula.

7. If A and B are formulas, then $(A \rightarrow B)$ is a formula.

8. If A is a formula and x is a variable, then $\forall x\, A$ is a formula.

9. If A is a formula and x is a variable, then $\exists x\, A$ is a formula.

10. Nothing else is a formula.

The definitions of the set of terms and that of formulas are *inductive definitions.* Essentially, we construct the set of formulas in infinitely many stages. In the initial stage, we pronounce all atomic formulas to be formulas; this corresponds to the first few cases of the definition, i.e., the cases for $\bot$, $R(t_1, \dots, t_n)$ and $=(t_1, t_2)$. "Atomic formula" thus means any formula of this form.

The other cases of the definition give rules for constructing new formulas out of formulas already constructed. At the second stage, we can use them to construct formulas out of atomic formulas. At the third stage, we construct new formulas from the atomic formulas and those obtained in the second stage, and so on. A formula is anything that is eventually constructed at such a stage, and nothing else.

By convention, we write $=$ between its arguments and leave out the parentheses: $t_1 = t_2$ is an abbreviation for $=(t_1, t_2)$. Moreover, $\neg{=}(t_1, t_2)$ is abbreviated as $t_1 \neq t_2$. When writing a formula $(B * C)$ constructed from B, C using a two-place connective $*$, we will often leave out the outermost pair of parentheses and write simply $B * C$.

Definition B.6. Formulas constructed using the defined operators are to be understood as follows:

1. $\top$ abbreviates $\neg\bot$.

2. $A \leftrightarrow B$ abbreviates $(A \rightarrow B) \land (B \rightarrow A)$.

If we work in a language for a specific application, we will often write two-place predicate symbols and function symbols between the respective terms, e.g., $t_1 < t_2$ and $(t_1 + t_2)$ in the language of arithmetic and $t_1 \in t_2$ in the language of set theory. The successor function in the language of arithmetic is even written conventionally *after* its argument: t'. Officially, however, these are just conventional abbreviations for $A^2_0(t_1, t_2)$, $f^2_0(t_1, t_2)$, $A^2_0(t_1, t_2)$ and $f^1_0(t)$, respectively.

Definition B.7 (Syntactic identity). The symbol $\equiv$ expresses syntactic identity between strings of symbols, i.e., $A \equiv B$ iff A and B are strings of symbols of the same length and which contain the same symbol in each place.

The $\equiv$ symbol may be flanked by strings obtained by concatenation, e.g., $A \equiv (B \lor C)$ means: the string of symbols A is the same string as the one obtained by concatenating an opening parenthesis, the string B, the $\lor$ symbol, the string C, and a closing parenthesis, in this order. If this is the case, then we know that the first symbol of A is an opening parenthesis, A contains B as a substring (starting at the second symbol), that substring is followed by $\lor$, etc.

B.3 Free Variables and Sentences

Definition B.8 (Free occurrences of a variable). The *free* occurrences of a variable in a formula are defined inductively as follows:

1. A is atomic: all variable occurrences in A are free.

2. $A \equiv \neg B$: the free variable occurrences of A are exactly those of B.

3. $A \equiv (B * C)$: the free variable occurrences of A are those in B together with those in C.

4. $A \equiv \forall x\, B$: the free variable occurrences in A are all of those in B except for occurrences of x.

5. $A \equiv \exists x\, B$: the free variable occurrences in A are all of those in B except for occurrences of x.

Definition B.9 (Bound Variables). An occurrence of a variable in a formula A is *bound* if it is not free.

Definition B.10 (Scope). If $\forall x\, B$ is an occurrence of a subformula in a formula A, then the corresponding occurrence of B in A is called the *scope* of the corresponding occurrence of $\forall x$. Similarly for $\exists x$.

If B is the scope of a quantifier occurrence $\forall x$ or $\exists x$ in A, then the free occurrences of x in B are bound in $\forall x\, B$ and $\exists x\, B$. We say that these occurrences are *bound by* the mentioned quantifier occurrence.

Example B.11. Consider the following formula:

$$\exists v_0 \underbrace{A^2_0(v_0, v_1)}_{B}$$

B represents the scope of $\exists v_0$. The quantifier binds the occurence of v_0 in B, but does not bind the occurence of v_1. So v_1 is a free variable in this case.

We can now see how this might work in a more complicated formula A:

$$\forall v_0 \underbrace{(A^1_0(v_0) \to A^2_0(v_0, v_1))}_{B} \to \exists v_1 \underbrace{(A^2_1(v_0, v_1) \lor \forall v_0 \overbrace{\neg A^1_1(v_0)}^{D})}_{C}$$

B is the scope of the first $\forall v_0$, C is the scope of $\exists v_1$, and D is the scope of the second $\forall v_0$. The first $\forall v_0$ binds the occurrences of v_0 in B, $\exists v_1$ the occurrence of v_1 in C, and the second $\forall v_0$ binds the occurrence of v_0 in D. The first occurrence of v_1 and the fourth occurrence of v_0 are free in A. The last occurrence of v_0 is free in D, but bound in C and A.

Definition B.12 (Sentence). A formula A is a *sentence* iff it contains no free occurrences of variables.

B.4 Substitution

Definition B.13 (Substitution in a term). We define $s[t/x]$, the result of *substituting* t for every occurrence of x in s, recursively:

1. $s \equiv c$: $s[t/x]$ is just s.
2. $s \equiv y$: $s[t/x]$ is also just s, provided y is a variable and $y \not\equiv x$.
3. $s \equiv x$: $s[t/x]$ is t.
4. $s \equiv f(t_1, \ldots, t_n)$: $s[t/x]$ is $f(t_1[t/x], \ldots, t_n[t/x])$.

Definition B.14. A term t is *free for* x in A if none of the free occurrences of x in A occur in the scope of a quantifier that binds a variable in t.

Example B.15.

1. v_8 is free for v_1 in $\exists v_3 A^2_4(v_3, v_1)$
2. $f^2_1(v_1, v_2)$ is *not* free for v_o in $\forall v_2 A^2_4(v_0, v_2)$

Definition B.16 (Substitution in a formula). If A is a formula, x is a variable, and t is a term free for x in A, then $A[t/x]$ is the result of substituting t for all free occurrences of x in A.

1. $A \equiv \bot$: $A[t/x]$ is $\bot$.
2. $A \equiv P(t_1, \ldots, t_n)$: $A[t/x]$ is $P(t_1[t/x], \ldots, t_n[t/x])$.

3. $A \equiv t_1 = t_2$: $A[t/x]$ is $t_1[t/x] = t_2[t/x]$.

4. $A \equiv \neg B$: $A[t/x]$ is $\neg B[t/x]$.

5. $A \equiv (B \wedge C)$: $A[t/x]$ is $(B[t/x] \wedge C[t/x])$.

6. $A \equiv (B \vee C)$: $A[t/x]$ is $(B[t/x] \vee C[t/x])$.

7. $A \equiv (B \rightarrow C)$: $A[t/x]$ is $(B[t/x] \rightarrow C[t/x])$.

8. $A \equiv \forall y\, B$: $A[t/x]$ is $\forall y\, B[t/x]$, provided y is a variable other than x; otherwise $A[t/x]$ is just A.

9. $A \equiv \exists y\, B$: $A[t/x]$ is $\exists y\, B[t/x]$, provided y is a variable other than x; otherwise $A[t/x]$ is just A.

Note that substitution may be vacuous: If x does not occur in A at all, then $A[t/x]$ is just A.

The restriction that t must be free for x in A is necessary to exclude cases like the following. If $A \equiv \exists y\, x < y$ and $t \equiv y$, then $A[t/x]$ would be $\exists y\, y < y$. In this case the free variable y is "captured" by the quantifier $\exists y$ upon substitution, and that is undesirable. For instance, we would like it to be the case that whenever $\forall x\, B$ holds, so does $B[t/x]$. But consider $\forall x\, \exists y\, x < y$ (here B is $\exists y\, x < y$). It is sentence that is true about, e.g., the natural numbers: for every number x there is a number y greater than it. If we allowed y as a possible substitution for x, we would end up with $B[y/x] \equiv \exists y\, y < y$, which is false. We prevent this by requiring that none of the free variables in t would end up being bound by a quantifier in A.

We often use the following convention to avoid cumbersume notation: If A is a formula with a free variable x, we write $A(x)$ to indicate this. When it is clear which A and x we have in mind, and t is a term (assumed to be free for x in $A(x)$), then we write $A(t)$ as short for $A(x)[t/x]$.

B.5 Structures for First-order Languages

First-order languages are, by themselves, *uninterpreted:* the constant symbols, function symbols, and predicate symbols have no specific meaning attached to them. Meanings are given by specifying a *structure.* It specifies the *domain*, i.e., the objects which the constant symbols pick out, the function symbols operate on, and the quantifiers range over. In addition, it specifies which constant symbols pick out which objects, how a function symbol maps objects to objects, and which objects the predicate symbols apply to. Structures are the basis for *semantic* notions in logic, e.g., the notion of consequence, validity, satisfiablity. They are variously called "structures," "interpretations," or "models" in the literature.

Definition B.17 (Structures). A *structure* M, for a language $\mathcal{L}$ of first-order logic consists of the following elements:

1. *Domain:* a non-empty set, $|M|$

2. *Interpretation of constant symbols:* for each constant symbol c of $\mathcal{L}$, an element $c^M \in |M|$

3. *Interpretation of predicate symbols:* for each n-place predicate symbol R of $\mathcal{L}$ (other than $=$), an n-place relation $R^M \subseteq |M|^n$

4. *Interpretation of function symbols:* for each n-place function symbol f of $\mathcal{L}$, an n-place function $f^M\colon |M|^n \to |M|$

Example B.18. A structure M for the language of arithmetic consists of a set, an element of $|M|$, 0^M, as interpretation of the constant symbol 0, a one-place function $\prime^M\colon |M| \to |M|$, two two-place functions $+^M$ and $\times^M$, both $|M|^2 \to |M|$, and a two-place relation $<^M \subseteq |M|^2$.

An obvious example of such a structure is the following:

1. $|N| = \mathbb{N}$

2. $\mathrm{o}^N = 0$

3. $\prime^N(n) = n + 1$ for all $n \in \mathbb{N}$

4. $+^N(n, m) = n + m$ for all $n, m \in \mathbb{N}$

5. $\times^N(n, m) = n \cdot m$ for all $n, m \in \mathbb{N}$

6. $<^N = \{\langle n, m \rangle : n \in \mathbb{N}, m \in \mathbb{N}, n < m\}$

The structure N for $\mathcal{L}_A$ so defined is called the *standard model of arithmetic*, because it interprets the non-logical constants of $\mathcal{L}_A$ exactly how you would expect.

However, there are many other possible structures for $\mathcal{L}_A$. For instance, we might take as the domain the set $\mathbb{Z}$ of integers instead of $\mathbb{N}$, and define the interpretations of o, $\prime$, $+$, $\times$, $<$ accordingly. But we can also define structures for $\mathcal{L}_A$ which have nothing even remotely to do with numbers.

Example B.19. A structure M for the language $\mathcal{L}_Z$ of set theory requires just a set and a single-two place relation. So technically, e.g., the set of people plus the relation "x is older than y" could be used as a structure for $\mathcal{L}_Z$, as well as $\mathbb{N}$ together with $n \geq m$ for $n, m \in \mathbb{N}$.

A particularly interesting structure for $\mathcal{L}_Z$ in which the elements of the domain are actually sets, and the interpretation of $\in$ actually is the relation "x is an element of y" is the structure HF of *hereditarily finite sets*:

1. $|HF| = \emptyset \cup \wp(\emptyset) \cup \wp(\wp(\emptyset)) \cup \wp(\wp(\wp(\emptyset))) \cup \dots;$

2. $\in^{HF} = \{\langle x, y \rangle : x, y \in |HF|, x \in y\}.$

The stipulations we make as to what counts as a structure impact our logic. For example, the choice to prevent empty domains ensures, given the usual account of satisfaction (or truth) for quantified sentences, that $\exists x\,(A(x) \lor \neg A(x))$ is valid—that is, a logical truth. And the stipulation that all constant symbols must refer to an object in the domain ensures that the existential generalization is a sound pattern of inference: $A(a)$, therefore $\exists x\, A(x)$. If we allowed names to refer outside the domain, or to not refer, then we would be on our way to a *free logic*, in which existential generalization requires an additional premise: $A(a)$ and $\exists x\, x = a$, therefore $\exists x\, A(x)$.

B.6 Satisfaction of a Formula in a Structure

The basic notion that relates expressions such as terms and formulas, on the one hand, and structures on the other, are those of ***value*** of a term and ***satisfaction*** of a formula. Informally, the value of a term is an element of a structure—if the term is just a constant, its value is the object assigned to the constant by the structure, and if it is built up using function symbols, the value is computed from the values of constants and the functions assigned to the functions in the term. A formula is *satisfied* in a structure if the interpretation given to the predicates makes the formula true in the domain of the structure. This notion of satisfaction is specified inductively: the specification of the structure directly states when atomic formulas are satisfied, and we define when a complex formula is satisfied depending on the main connective or quantifier and whether or not the immediate subformulas are satisfied. The case of the quantifiers here is a bit tricky, as the immediate subformula of a quantified formula has a free variable, and structures don't specify the values of variables. In order to deal with this difficulty, we also introduce ***variable assignments*** and define satisfaction not with respect to a structure alone, but with respect to a structure plus a variable assignment.

Definition B.20 (Variable Assignment). A *variable assignment* s for a structure M is a function which maps each variable to an element of $|M|$, i.e., $s\colon \mathrm{Var} \to |M|$.

A structure assigns a value to each constant symbol, and a variable assignment to each variable. But we want to use terms built up from them to also name elements of the domain. For this we define the value of terms inductively. For constant symbols and variables the value is just as the structure or the variable assignment specifies it; for more complex terms it is computed recursively using the functions the structure assigns to the function symbols.

Definition B.21 (Value of Terms). If t is a term of the language $\mathcal{L}$, M is a structure for $\mathcal{L}$, and s is a variable assignment for M, the *value* $\mathrm{Val}_s^M(t)$ is defined as follows:

1. $t \equiv c$: $\mathrm{Val}_s^M(t) = c^M$.
2. $t \equiv x$: $\mathrm{Val}_s^M(t) = s(x)$.
3. $t \equiv f(t_1, \dots, t_n)$:

$$\mathrm{Val}_s^M(t) = f^M(\mathrm{Val}_s^M(t_1), \dots, \mathrm{Val}_s^M(t_n)).$$

Definition B.22 (x-Variant). If s is a variable assignment for a structure M, then any variable assignment s' for M which differs from s at most in what it assigns to x is called an *x-variant* of s. If s' is an x-variant of s we write $s \sim_x s'$.

Note that an x-variant of an assignment s does not *have* to assign something different to x. In fact, every assignment counts as an x-variant of itself.

Definition B.23 (Satisfaction). Satisfaction of a formula A in a structure M relative to a variable assignment s, in symbols: $M, s \vDash A$, is defined recursively as follows. (We write $M, s \nvDash A$ to mean "not $M, s \vDash A$.")

1. $A \equiv \bot$: $M, s \nvDash A$.
2. $A \equiv R(t_1, \ldots, t_n)$: $M, s \vDash A$ iff $\langle \mathrm{Val}_s^M(t_1), \ldots, \mathrm{Val}_s^M(t_n) \rangle \in R^M$.
3. $A \equiv t_1 = t_2$: $M, s \vDash A$ iff $\mathrm{Val}_s^M(t_1) = \mathrm{Val}_s^M(t_2)$.
4. $A \equiv \neg B$: $M, s \vDash A$ iff $M, s \nvDash B$.
5. $A \equiv (B \land C)$: $M, s \vDash A$ iff $M, s \vDash B$ and $M, s \vDash C$.
6. $A \equiv (B \lor C)$: $M, s \vDash A$ iff $M, s \vDash A$ or $M, s \vDash B$ (or both).
7. $A \equiv (B \to C)$: $M, s \vDash A$ iff $M, s \nvDash B$ or $M, s \vDash C$ (or both).
8. $A \equiv \forall x\, B$: $M, s \vDash A$ iff for every x-variant s' of s, $M, s' \vDash B$.
9. $A \equiv \exists x\, B$: $M, s \vDash A$ iff there is an x-variant s' of s so that $M, s' \vDash B$.

The variable assignments are important in the last two clauses. We cannot define satisfaction of $\forall x\, B(x)$ by "for all $a \in |M|$, $M \vDash B(a)$." We cannot define satisfaction of $\exists x\, B(x)$ by "for at least one $a \in |M|$, $M \vDash B(a)$." The reason is that a is not symbol of the language, and so $B(a)$ is not a formula (that is, $B[a/x]$ is undefined). We also cannot assume that we have constant symbols or terms available that name every element of M, since there is nothing in the definition of structures that requires it. Even in the standard language the set of constant symbols is countably infinite, so if $|M|$ is not countable there aren't even enough constant symbols to name every object.

Example B.24. Let $\mathcal{L} = \{a, b, f, R\}$ where a and b are constant symbols, f is a two-place function symbol, and R is a two-place predicate symbol. Consider the structure $\mathbf{M}$ defined by:

1. $|\mathbf{M}| = \{1, 2, 3, 4\}$
2. $a^{\mathbf{M}} = 1$
3. $b^{\mathbf{M}} = 2$
4. $f^{\mathbf{M}}(x, y) = x + y$ if $x + y \leq 3$ and $= 3$ otherwise.
5. $R^{\mathbf{M}} = \{\langle 1, 1\rangle, \langle 1, 2\rangle, \langle 2, 3\rangle, \langle 2, 4\rangle\}$

The function $s(x) = 1$ that assigns $1 \in |\mathbf{M}|$ to every variable is a variable assignment for $\mathbf{M}$.

Then

$$\mathrm{Val}_s^{\mathbf{M}}(f(a, b)) = f^{\mathbf{M}}(\mathrm{Val}_s^{\mathbf{M}}(a), \mathrm{Val}_s^{\mathbf{M}}(b)).$$

Since a and b are constant symbols, $\mathrm{Val}_s^{\mathbf{M}}(a) = a^{\mathbf{M}} = 1$ and $\mathrm{Val}_s^{\mathbf{M}}(b) = b^{\mathbf{M}} = 2$. So

$$\mathrm{Val}_s^{\mathbf{M}}(f(a, b)) = f^{\mathbf{M}}(1, 2) = 1 + 2 = 3.$$

To compute the value of $f(f(a, b), a)$ we have to consider

$$\mathrm{Val}_s^{\mathbf{M}}(f(f(a, b), a)) = f^{\mathbf{M}}(\mathrm{Val}_s^{\mathbf{M}}(f(a, b)), \mathrm{Val}_s^{\mathbf{M}}(a)) = f^{\mathbf{M}}(3, 1) = 3,$$

since $3 + 1 > 3$. Since $s(x) = 1$ and $\mathrm{Val}_s^{\mathbf{M}}(x) = s(x)$, we also have

$$\mathrm{Val}_s^{\mathbf{M}}(f(f(a, b), x)) = f^{\mathbf{M}}(\mathrm{Val}_s^{\mathbf{M}}(f(a, b)), \mathrm{Val}_s^{\mathbf{M}}(x)) = f^{\mathbf{M}}(3, 1) = 3,$$

An atomic formula $R(t_1, t_2)$ is satisfied if the tuple of values of its arguments, i.e., $\langle \mathrm{Val}_s^{\mathbf{M}}(t_1), \mathrm{Val}_s^{\mathbf{M}}(t_2)\rangle$, is an element of $R^{\mathbf{M}}$. So, e.g., we have $\mathbf{M}, s \vDash R(b, f(a, b))$ since $\langle \mathrm{Val}^{\mathbf{M}}(b), \mathrm{Val}^{\mathbf{M}}(f(a, b))\rangle = \langle 2, 3\rangle \in R^{\mathbf{M}}$, but $\mathbf{M}, s \nvDash R(x, f(a, b))$ since $\langle 1, 3\rangle \notin R^{\mathbf{M}}[s]$.

To determine if a non-atomic formula A is satisfied, you apply the clauses in the inductive definition that applies to the main connective. For instance, the main connective in $R(a, a) \rightarrow (R(b, x) \lor R(x, b)$ is the $\rightarrow$, and

$$\mathbf{M}, s \vDash R(a, a) \rightarrow (R(b, x) \lor R(x, b)) \text{ iff}$$

$$\boldsymbol{M}, s \nvDash R(a, a) \text{ or } \boldsymbol{M}, s \vDash R(b, x) \lor R(x, b)$$

Since $\boldsymbol{M}, s \vDash R(a, a)$ (because $\langle 1, 1 \rangle \in R^{\boldsymbol{M}}$) we can't yet determine the answer and must first figure out if $\boldsymbol{M}, s \vDash R(b, x) \lor R(x, b)$:

$$\begin{aligned} &\boldsymbol{M}, s \vDash R(b, x) \lor R(x, b) \text{ iff} \\ &\qquad \boldsymbol{M}, s \vDash R(b, x) \text{ or } \boldsymbol{M}, s \vDash R(x, b) \end{aligned}$$

And this is the case, since $\boldsymbol{M}, s \vDash R(x, b)$ (because $\langle 1, 2 \rangle \in R^{\boldsymbol{M}}$).

Recall that an x-variant of s is a variable assignment that differs from s at most in what it assigns to x. For every element of $|\boldsymbol{M}|$, there is an x-variant of s: $s_1(x) = 1$, $s_2(x) = 2$, $s_3(x) = 3$, $s_4(x) = 4$, and with $s_i(y) = s(y) = 1$ for all variables y other than x. These are all the x-variants of s for the structure $\boldsymbol{M}$, since $|\boldsymbol{M}| = \{1, 2, 3, 4\}$. Note, in particular, that $s_1 = s$ is also an x-variant of s, i.e., s is always an x-variant of itself.

To determine if an existentially quantified formula $\exists x\, A(x)$ is satisfied, we have to determine if $\boldsymbol{M}, s' \vDash A(x)$ for at least one x-variant s' of s. So,

$$\boldsymbol{M}, s \vDash \exists x\, (R(b, x) \lor R(x, b)),$$

since $\boldsymbol{M}, s_1 \vDash R(b, x) \lor R(x, b)$ (s_3 would also fit the bill). But,

$$\boldsymbol{M}, s \nvDash \exists x\, (R(b, x) \land R(x, b))$$

since for none of the s_i, $\boldsymbol{M}, s_i \vDash R(b, x) \land R(x, b)$.

To determine if a universally quantified formula $\forall x\, A(x)$ is satisfied, we have to determine if $\boldsymbol{M}, s' \vDash A(x)$ for all x-variants s' of s. So,

$$\boldsymbol{M}, s \vDash \forall x\, (R(x, a) \rightarrow R(a, x)),$$

since $\boldsymbol{M}, s_i \vDash R(x, a) \rightarrow R(a, x)$ for all s_i ($\boldsymbol{M}, s_1 \vDash R(a, x)$ and $\boldsymbol{M}, s_j \nvDash R(x, a)$ for $j = 2$, 3, and 4). But,

$$\boldsymbol{M}, s \nvDash \forall x\, (R(a, x) \rightarrow R(x, a))$$

since $M, s_2 \nvDash R(a, x) \rightarrow R(x, a)$ (because $M, s_2 \vDash R(a, x)$ and $M, s_2 \nvDash R(x, a)$).

For a more complicated case, consider

$$\forall x\,(R(a, x) \rightarrow \exists y\, R(x, y)).$$

Since $M, s_3 \nvDash R(a, x)$ and $M, s_4 \nvDash R(a, x)$, the interesting cases where we have to worry about the consequent of the conditional are only s_1 and s_2. Does $M, s_1 \vDash \exists y\, R(x, y)$ hold? It does if there is at least one y-variant s_1' of s_1 so that $M, s_1' \vDash R(x, y)$. In fact, s_1 is such a y-variant ($s_1(x) = 1$, $s_1(y) = 1$, and $\langle 1, 1\rangle \in R^M$), so the answer is yes. To determine if $M, s_2 \vDash \exists y\, R(x, y)$ we have to look at the y-variants of s_2. Here, s_2 itself does not satisfy $R(x, y)$ ($s_2(x) = 2$, $s_2(y) = 1$, and $\langle 2, 1\rangle \notin R^M$). However, consider $s_2' \sim_y s_2$ with $s_2'(y) = 3$. $M, s_2' \vDash R(x, y)$ since $\langle 2, 3\rangle \in R^M$, and so $M, s_2 \vDash \exists y\, R(x, y)$. In sum, for every x-variant s_i of s, either $M, s_i \nvDash R(a, x)$ ($i = 3$, 4) or $M, s_i \vDash \exists y\, R(x, y)$ ($i = 1$, 2), and so

$$M, s \vDash \forall x\,(R(a, x) \rightarrow \exists y\, R(x, y)).$$

On the other hand,

$$M, s \nvDash \exists x\,(R(a, x) \land \forall y\, R(x, y)).$$

The only x-variants s_i of s with $M, s_i \vDash R(a, x)$ are s_1 and s_2. But for each, there is in turn a y-variant $s_i' \sim_y s_i$ with $s_i'(y) = 4$ so that $M, s_i' \nvDash R(x, y)$ and so $M, s_i \nvDash \forall y\, R(x, y)$ for $i = 1$, 2. In sum, none of the x-variants $s_i \sim_x s$ are such that $M, s_i \vDash R(a, x) \land \forall y\, R(x, y)$.

B.7 Variable Assignments

A variable assignment s provides a value for *every* variable—and there are infinitely many of them. This is of course not necessary. We require variable assignments to assign values to all variables simply because it makes things a lot easier. The value of a term t, and whether or not a formula A is satisfied in a structure with respect to s, only depend on the assignments s makes to the variables in t and the free variables of A. This is the content of the next two propositions. To make the idea of "depends on" precise, we show that any two variable assignments that agree on all the variables in t give the same value, and that A is satisfied relative to one iff it is satisfied relative to the other if two variable assignments agree on all free variables of A.

Proposition B.25. *If the variables in a term t are among $x_1, \dots, x_n$, and $s_1(x_i) = s_2(x_i)$ for $i = 1, \dots, n$, then* $\mathrm{Val}^{\mathbf{M}}_{s_1}(t) = \mathrm{Val}^{\mathbf{M}}_{s_2}(t)$.

Proof. By induction on the complexity of t. For the base case, t can be a constant symbol or one of the variables $x_1, \dots, x_n$. If $t = c$, then $\mathrm{Val}^{\mathbf{M}}_{s_1}(t) = c^{\mathbf{M}} = \mathrm{Val}^{\mathbf{M}}_{s_2}(t)$. If $t = x_i$, $s_1(x_i) = s_2(x_i)$ by the hypothesis of the proposition, and so $\mathrm{Val}^{\mathbf{M}}_{s_1}(t) = s_1(x_i) = s_2(x_i) = \mathrm{Val}^{\mathbf{M}}_{s_2}(t)$.

For the inductive step, assume that $t = f(t_1, \dots, t_k)$ and that the claim holds for $t_1, \dots, t_k$. Then

$$\begin{aligned}\mathrm{Val}^{\mathbf{M}}_{s_1}(t) &= \mathrm{Val}^{\mathbf{M}}_{s_1}(f(t_1, \dots, t_k)) = \\ &= f^{\mathbf{M}}(\mathrm{Val}^{\mathbf{M}}_{s_1}(t_1), \dots, \mathrm{Val}^{\mathbf{M}}_{s_1}(t_k))\end{aligned}$$

For $j = 1, \dots, k$, the variables of t_j are among $x_1, \dots, x_n$. So by induction hypothesis, $\mathrm{Val}^{\mathbf{M}}_{s_1}(t_j) = \mathrm{Val}^{\mathbf{M}}_{s_2}(t_j)$. So,

$$\begin{aligned}\mathrm{Val}^{\mathbf{M}}_{s_1}(t) &= \mathrm{Val}^{\mathbf{M}}_{s_2}(f(t_1, \dots, t_k)) = \\ &= f^{\mathbf{M}}(\mathrm{Val}^{\mathbf{M}}_{s_1}(t_1), \dots, \mathrm{Val}^{\mathbf{M}}_{s_1}(t_k)) = \\ &= f^{\mathbf{M}}(\mathrm{Val}^{\mathbf{M}}_{s_2}(t_1), \dots, \mathrm{Val}^{\mathbf{M}}_{s_2}(t_k)) = \\ &= \mathrm{Val}^{\mathbf{M}}_{s_2}(f(t_1, \dots, t_k)) = \mathrm{Val}^{\mathbf{M}}_{s_2}(t).\end{aligned}$$

□

Proposition B.26. *If the free variables in A are among $x_1, \dots, x_n$, and $s_1(x_i) = s_2(x_i)$ for $i = 1, \dots, n$, then $\mathbf{M}, s_1 \vDash A$ iff $\mathbf{M}, s_2 \vDash A$.*

Proof. We use induction on the complexity of A. For the base case, where A is atomic, A can be: $\bot$, $R(t_1, \dots, t_k)$ for a k-place predicate R and terms $t_1, \dots, t_k$, or $t_1 = t_2$ for terms t_1 and t_2.

1. $A \equiv \bot$: both $M, s_1 \nvDash A$ and $M, s_2 \nvDash A$.

2. $A \equiv R(t_1, \ldots, t_k)$: let $M, s_1 \vDash A$. Then

$$\langle \mathrm{Val}^M_{s_1}(t_1), \ldots, \mathrm{Val}^M_{s_1}(t_k)\rangle \in R^M.$$

   For $i = 1, \ldots, k$, $\mathrm{Val}^M_{s_1}(t_i) = \mathrm{Val}^M_{s_2}(t_i)$ by Proposition B.25. So we also have $\langle \mathrm{Val}^M_{s_2}(t_i), \ldots, \mathrm{Val}^M_{s_2}(t_k)\rangle \in R^M$.

3. $A \equiv t_1 = t_2$: suppose $M, s_1 \vDash A$. Then $\mathrm{Val}^M_{s_1}(t_1) = \mathrm{Val}^M_{s_1}(t_2)$. So,

$$\begin{aligned} \mathrm{Val}^M_{s_2}(t_1) &= \mathrm{Val}^M_{s_1}(t_1) && \text{(by Proposition B.25)} \\ &= \mathrm{Val}^M_{s_1}(t_2) && \text{(since } M, s_1 \vDash t_1 = t_2\text{)} \\ &= \mathrm{Val}^M_{s_2}(t_2) && \text{(by Proposition B.25),} \end{aligned}$$

   so $M, s_2 \vDash t_1 = t_2$.

Now assume $M, s_1 \vDash B$ iff $M, s_2 \vDash B$ for all formulas B less complex than A. The induction step proceeds by cases determined by the main operator of A. In each case, we only demonstrate the forward direction of the biconditional; the proof of the reverse direction is symmetrical. In all cases except those for the quantifiers, we apply the induction hypothesis to sub-formulas B of A. The free variables of B are among those of A. Thus, if s_1 and s_2 agree on the free variables of A, they also agree on those of B, and the induction hypothesis applies to B.

1. $A \equiv \neg B$: if $M, s_1 \vDash A$, then $M, s_1 \nvDash B$, so by the induction hypothesis, $M, s_2 \nvDash B$, hence $M, s_2 \vDash A$.

2. $A \equiv B \land C$: exercise.

3. $A \equiv B \lor C$: if $M, s_1 \vDash A$, then $M, s_1 \vDash B$ or $M, s_1 \vDash C$. By induction hypothesis, $M, s_2 \vDash B$ or $M, s_2 \vDash C$, so $M, s_2 \vDash A$.

4. $A \equiv B \to C$: exercise.

5. $A \equiv \exists x\, B$: if $M, s_1 \vDash A$, there is an x-variant s_1' of s_1 so that $M, s_1' \vDash B$. Let s_2' be the x-variant of s_2 that assigns the same thing to x as does s_1'. The free variables of B are among $x_1, \ldots, x_n$, and x. $s_1'(x_i) = s_2'(x_i)$, since s_1' and s_2' are x-variants of s_1 and s_2, respectively, and by hypothesis $s_1(x_i) = s_2(x_i)$. $s_1'(x) = s_2'(x)$ by the way we have defined s_2'. Then the induction hypothesis applies to B and s_1', s_2', so $M, s_2' \vDash B$. Hence, there is an x-variant of s_2 that satisfies B, and so $M, s_2 \vDash A$.

6. $A \equiv \forall x\, B$: exercise.

By induction, we get that $M, s_1 \vDash A$ iff $M, s_2 \vDash A$ whenever the free variables in A are among $x_1, \ldots, x_n$ and $s_1(x_i) = s_2(x_i)$ for $i = 1, \ldots, n$. □

Sentences have no free variables, so any two variable assignments assign the same things to all the (zero) free variables of any sentence. The proposition just proved then means that whether or not a sentence is satisfied in a structure relative to a variable assignment is completely independent of the assignment. We'll record this fact. It justifies the definition of satisfaction of a sentence in a structure (without mentioning a variable assignment) that follows.

Corollary B.27. *If A is a sentence and s a variable assignment, then $M, s \vDash A$ iff $M, s' \vDash A$ for every variable assignment s'.*

Proof. Let s' be any variable assignment. Since A is a sentence, it has no free variables, and so every variable assignment s' trivially assigns the same things to all free variables of A as does s. So the condition of Proposition B.26 is satisfied, and we have $M, s \vDash A$ iff $M, s' \vDash A$. □

Definition B.28. If A is a sentence, we say that a structure M *satisfies* A, $M \vDash A$, iff $M, s \vDash A$ for all variable assignments s.

If $M \vDash A$, we also simply say that A *is true in* M.

Proposition B.29. *Let M be a structure, A be a sentence, and s a variable assignment. $M \vDash A$ iff $M, s \vDash A$.*

Proof. Exercise. □

Proposition B.30. *Suppose $A(x)$ only contains x free, and M is a structure. Then:*

1. *$M \vDash \exists x\, A(x)$ iff $M, s \vDash A(x)$ for at least one variable assignment s.*

2. *$M \vDash \forall x\, A(x)$ iff $M, s \vDash A(x)$ for all variable assignments s.*

Proof. Exercise. □

B.8 Extensionality

Extensionality, sometimes called relevance, can be expressed informally as follows: the only factors that bears upon the satisfaction of formula A in a structure M relative to a variable assignment s, are the size of the domain and the assignments made by M and s to the elements of the language that actually appear in A.

One immediate consequence of extensionality is that where two structures M and M' agree on all the elements of the language appearing in a sentence A and have the same domain, M and M' must also agree on whether or not A itself is true.

Proposition B.31 (Extensionality). *Let A be a formula, and M_1 and M_2 be structures with $|M_1| = |M_2|$, and s a variable assignment on $|M_1| = |M_2|$. If $c^{M_1} = c^{M_2}$, $R^{M_1} = R^{M_2}$, and $f^{M_1} = f^{M_2}$ for every* *constant symbol c, relation symbol R, and function symbol f occurring in A, then $\mathbf{M}_1, s \vDash A$ iff $\mathbf{M}_2, s \vDash A$.*

Proof. First prove (by induction on t) that for every term, $\mathrm{Val}_s^{M_1}(t) = \mathrm{Val}_s^{M_2}(t)$. Then prove the proposition by induction on A, making use of the claim just proved for the induction basis (where A is atomic). □

Corollary B.32 (Extensionality for Sentences). *Let A be a sentence and $\mathbf{M}_1$, $\mathbf{M}_2$ as in Proposition B.31. Then $\mathbf{M}_1 \vDash A$ iff $\mathbf{M}_2 \vDash A$.*

Proof. Follows from Proposition B.31 by Corollary B.27. □

Moreover, the value of a term, and whether or not a structure satisfies a formula, only depends on the values of its subterms.

Proposition B.33. *Let $\mathbf{M}$ be a structure, t and t' terms, and s a variable assignment. Let $s' \sim_x s$ be the x-variant of s given by $s'(x) =$ $\mathrm{Val}_s^{\mathbf{M}}(t')$. Then $\mathrm{Val}_s^{\mathbf{M}}(t[t'/x]) = \mathrm{Val}_{s'}^{\mathbf{M}}(t)$.*

Proof. By induction on t.

1. If t is a constant, say, $t \equiv c$, then $t[t'/x] = c$, and $\mathrm{Val}_s^{\mathbf{M}}(c) = c^{\mathbf{M}} = \mathrm{Val}_{s'}^{\mathbf{M}}(c)$.

2. If t is a variable other than x, say, $t \equiv y$, then $t[t'/x] = y$, and $\mathrm{Val}_s^{\mathbf{M}}(y) = \mathrm{Val}_{s'}^{\mathbf{M}}(y)$ since $s' \sim_x s$.

3. If $t \equiv x$, then $t[t'/x] = t'$. But $\mathrm{Val}_{s'}^{\mathbf{M}}(x) = \mathrm{Val}_s^{\mathbf{M}}(t')$ by definition of s'.

4. If $t \equiv f(t_1, \ldots, t_n)$ then we have:

$$\mathrm{Val}_s^{\mathbf{M}}(t[t'/x]) =$$

$$= \mathrm{Val}^{M}_{s}(f(t_1[t'/x], \dots, t_n[t'/x]))$$
by definition of $t[t'/x]$
$$= f^{M}(\mathrm{Val}^{M}_{s}(t_1[t'/x]), \dots, \mathrm{Val}^{M}_{s}(t_n[t'/x]))$$
by definition of $\mathrm{Val}^{M}_{s}(f(\dots))$
$$= f^{M}(\mathrm{Val}^{M}_{s'}(t_1), \dots, \mathrm{Val}^{M}_{s'}(t_n))$$
by induction hypothesis
$$= \mathrm{Val}^{M}_{s'}(t) \text{ by definition of } \mathrm{Val}^{M}_{s'}(f(\dots))$$ □

Proposition B.34. *Let M be a structure, A a formula, t a term, and s a variable assignment. Let $s' \sim_x s$ be the x-variant of s given by $s'(x) = \mathrm{Val}^{M}_{s}(t)$. Then $M, s \vDash A[t/x]$ iff $M, s' \vDash A$.*

Proof. Exercise. □

B.9 Semantic Notions

Give the definition of structures for first-order languages, we can define some basic semantic properties of and relationships between sentences. The simplest of these is the notion of *validity* of a sentence. A sentence is valid if it is satisfied in every structure. Valid sentences are those that are satisfied regardless of how the non-logical symbols in it are interpreted. Valid sentences are therefore also called *logical truths*—they are true, i.e., satisfied, in any structure and hence their truth depends only on the logical symbols occurring in them and their syntactic structure, but not on the non-logical symbols or their interpretation.

Definition B.35 (Validity). A sentence A is *valid*, $\vDash A$, iff $M \vDash A$ for every structure M.

Definition B.36 (Entailment). A set of sentences Γ *entails* a sentence A, $\Gamma \vDash A$, iff for every structure $\mathbf{M}$ with $\mathbf{M} \vDash \Gamma$, $\mathbf{M} \vDash A$.

Definition B.37 (Satisfiability). A set of sentences Γ is *satisfiable* if $\mathbf{M} \vDash \Gamma$ for some structure $\mathbf{M}$. If Γ is not satisfiable it is called *unsatisfiable.*

Proposition B.38. *A sentence A is valid iff $\Gamma \vDash A$ for every set of sentences Γ.*

Proof. For the forward direction, let A be valid, and let Γ be a set of sentences. Let $\mathbf{M}$ be a structure so that $\mathbf{M} \vDash \Gamma$. Since A is valid, $\mathbf{M} \vDash A$, hence $\Gamma \vDash A$.

For the contrapositive of the reverse direction, let A be invalid, so there is a structure $\mathbf{M}$ with $\mathbf{M} \nvDash A$. When $\Gamma = \{\top\}$, since $\top$ is valid, $\mathbf{M} \vDash \Gamma$. Hence, there is a structure $\mathbf{M}$ so that $\mathbf{M} \vDash \Gamma$ but $\mathbf{M} \nvDash A$, hence Γ does not entail A. □

Proposition B.39. *$\Gamma \vDash A$ iff $\Gamma \cup \{\neg A\}$ is unsatisfiable.*

Proof. For the forward direction, suppose $\Gamma \vDash A$ and suppose to the contrary that there is a structure $\mathbf{M}$ so that $\mathbf{M} \vDash \Gamma \cup \{\neg A\}$. Since $\mathbf{M} \vDash \Gamma$ and $\Gamma \vDash A$, $\mathbf{M} \vDash A$. Also, since $\mathbf{M} \vDash \Gamma \cup \{\neg A\}$, $\mathbf{M} \vDash \neg A$, so we have both $\mathbf{M} \vDash A$ and $\mathbf{M} \nvDash A$, a contradiction. Hence, there can be no such structure $\mathbf{M}$, so $\Gamma \cup \{A\}$ is unsatisfiable.

For the reverse direction, suppose $\Gamma \cup \{\neg A\}$ is unsatisfiable. So for every structure $\mathbf{M}$, either $\mathbf{M} \nvDash \Gamma$ or $\mathbf{M} \vDash A$. Hence, for every structure $\mathbf{M}$ with $\mathbf{M} \vDash \Gamma$, $\mathbf{M} \vDash A$, so $\Gamma \vDash A$. □

Proposition B.40. *If $\Gamma \subseteq \Gamma'$ and $\Gamma \vDash A$, then $\Gamma' \vDash A$.*

Proof. Suppose that $\Gamma \subseteq \Gamma'$ and $\Gamma \vDash A$. Let $\mathbf{M}$ be such that $\mathbf{M} \vDash \Gamma'$; then $\mathbf{M} \vDash \Gamma$, and since $\Gamma \vDash A$, we get that $\mathbf{M} \vDash A$. Hence, whenever $\mathbf{M} \vDash \Gamma'$, $\mathbf{M} \vDash A$, so $\Gamma' \vDash A$. □

Theorem B.41 (Semantic Deduction Theorem). $\Gamma \cup \{A\} \vDash B$ *iff* $\Gamma \vDash A \rightarrow B$.

Proof. For the forward direction, let $\Gamma \cup \{A\} \vDash B$ and let $\mathbf{M}$ be a structure so that $\mathbf{M} \vDash \Gamma$. If $\mathbf{M} \vDash A$, then $\mathbf{M} \vDash \Gamma \cup \{A\}$, so since $\Gamma \cup \{A\}$ entails B, we get $\mathbf{M} \vDash B$. Therefore, $\mathbf{M} \vDash A \rightarrow B$, so $\Gamma \vDash A \rightarrow B$.

For the reverse direction, let $\Gamma \vDash A \rightarrow B$ and $\mathbf{M}$ be a structure so that $\mathbf{M} \vDash \Gamma \cup \{A\}$. Then $\mathbf{M} \vDash \Gamma$, so $\mathbf{M} \vDash A \rightarrow B$, and since $\mathbf{M} \vDash A$, $\mathbf{M} \vDash B$. Hence, whenever $\mathbf{M} \vDash \Gamma \cup \{A\}$, $\mathbf{M} \vDash B$, so $\Gamma \cup \{A\} \vDash B$. □

Proposition B.42. *Let $\mathbf{M}$ be a structure, and $A(x)$ a formula with one free variable x, and t a closed term. Then:*

1. $A(t) \vDash \exists x\, A(x)$
2. $\forall x\, A(x) \vDash A(t)$

Proof.
1. Suppose $\mathbf{M} \vDash A(t)$. Let s be a variable assignment with $s(x) = \mathrm{Val}^{\mathbf{M}}(t)$. Then $\mathbf{M}, s \vDash A(t)$ since $A(t)$ is a sentence. By Proposition B.34, $\mathbf{M}, s \vDash A(x)$. By Proposition B.30, $\mathbf{M} \vDash \exists x\, A(x)$.

2. Exercise. □

B.10 Theories

Definition B.43. A set of sentences Γ is *closed* iff, whenever $\Gamma \vDash A$ then $A \in \Gamma$. The *closure* of a set of sentences Γ is $\{A : \Gamma \vDash A\}$.

We say that Γ is ***axiomatized by*** a set of sentences Δ if Γ is the closure of Δ

Example B.44. The theory of strict linear orders in the language $\mathcal{L}_<$ is axiomatized by the set

$$\forall x\, \neg x < x,$$
$$\forall x\, \forall y\, ((x < y \lor y < x) \lor x = y),$$
$$\forall x\, \forall y\, \forall z\, ((x < y \land y < z) \to x < z)$$

It completely captures the intended structures: every strict linear order is a model of this axiom system, and vice versa, if R is a linear order on a set X, then the structure M with $|M| = X$ and $<^M = R$ is a model of this theory.

Example B.45. The theory of groups in the language 1 (constant symbol), $\cdot$ (two-place function symbol) is axiomatized by

$$\forall x\, (x \cdot 1) = x$$
$$\forall x\, \forall y\, \forall z\, (x \cdot (y \cdot z)) = ((x \cdot y) \cdot z)$$
$$\forall x\, \exists y\, (x \cdot y) = 1$$

Example B.46. The theory of Peano arithmetic is axiomatized by the following sentences in the language of arithmetic $\mathcal{L}_A$.

$$\neg \exists x\, x' = 0$$
$$\forall x\, \forall y\, (x' = y' \to x = y)$$
$$\forall x\, \forall y\, (x < y \leftrightarrow \exists z\, (z' + x) = y)$$
$$\forall x\, (x + 0) = x$$
$$\forall x\, \forall y\, (x + y') = (x + y)'$$
$$\forall x\, (x \times 0) = 0$$

$$\forall x \, \forall y \, (x \times y') = ((x \times y) + x)$$

plus all sentences of the form

$$(A(0) \land \forall x \, (A(x) \rightarrow A(x'))) \rightarrow \forall x \, A(x)$$

Since there are infinitely many sentences of the latter form, this axiom system is infinite. The latter form is called the *induction schema*. (Actually, the induction schema is a bit more complicated than we let on here.)

The third axiom is an *explicit definition* of $<$.

Summary

A **first-order language** consists of **constant**, **function**, and **predicate** symbols. Function and constant symbols take a specified number of arguments. In the **language of arithmetic**, e.g., we have a single constant symbol 0, one 1-place function symbol $'$, two 2-place function symbols $+$ and $\times$, and one 2-place predicate symbol $<$. From **variables** and constant and function symbols we form the **terms** of a language. From the terms of a language together with its predicate symbol, as well as the **identity symbol** $=$, we form the **atomic formulas**. And in turn from them, using the logical connectives $\lnot$, $\lor$, $\land$, $\rightarrow$, $\leftrightarrow$ and the quantifiers $\forall$ and $\exists$ we form its formulas. Since we are careful to always include necessary parentheses in the process of forming terms and formulas, there is always exactly one way of reading a formula. This makes it possible to define things by induction on the structure of formulas.

Occurrences of variables in formulas are sometimes governed by a corresponding quantifier: if a variable occurs in the **scope** of a quantifier it is considered **bound**, otherwise **free**. These concepts all have inductive definitions, and we also inductively define the operation of **substitution** of a term for a variable in a formula. Formulas without free variable occurrences are called **sentences**.

The **semantics** for a first-order language is given by a **structure** for that language. It consists of a **domain** and elements of that domain are assigned to each constant symbol. Function symbols are interpreted by functions and relation symbols by relation on the domain. A function from the set of variables to the domain is a **variable assignment**. The relation of **satisfaction** relates structures, variable assignments and formulas; $M, s \vDash A$ is defined by induction on the structure of A. $M, s \vDash A$ only depends on the interpretation of the symbols actually occurring in A, and in particular does not depend on s if A contains no free variables. So if A is a sentence, $M \vDash A$ if $M, s \vDash A$ for any (or all) s.

The satisfaction relation is the basis for all semantic notions. A sentence is **valid**, $\vDash A$, if it is satisfied in every structure. A sentence A is **entailed** by set of sentences Γ, $\Gamma \vDash A$, iff $M \vDash A$ for all M which satisfy every sentence in Γ. A set Γ is **satisfiable** iff there is some structure that satisfies every sentence in Γ, otherwise unsatisfiable. These notions are interrelated, e.g., $\Gamma \vDash A$ iff $\Gamma \cup \{\neg A\}$ is unsatisfiable.

Problems

Problem B.1. Give an inductive definition of the bound variable occurrences along the lines of Definition B.8.

Problem B.2. Let $\mathcal{L} = \{c, f, A\}$ with one constant symbol, one one-place function symbol and one two-place predicate symbol, and let the structure M be given by

1. $|M| = \{1, 2, 3\}$

2. $c^M = 3$

3. $f^M(1) = 2, f^M(2) = 3, f^M(3) = 2$

4. $A^M = \{\langle 1, 2\rangle, \langle 2, 3\rangle, \langle 3, 3\rangle\}$

(a) Let $s(v) = 1$ for all variables v. Find out whether

$$M, s \vDash \exists x\,(A(f(z), c) \rightarrow \forall y\,(A(y, x) \vee A(f(y), x)))$$

Explain why or why not.

(b) Give a different structure and variable assignment in which the formula is not satisfied.

Problem B.3. Complete the proof of Proposition B.26.

Problem B.4. Prove Proposition B.29

Problem B.5. Prove Proposition B.30.

Problem B.6. Suppose $\mathcal{L}$ is a language without function symbols. Given a structure M, c a constant symbol and $a \in |M|$, define $M[a/c]$ to be the structure that is just like M, except that $c^{M[a/c]} = a$. Define $M \Vdash A$ for sentences A by:

1. $A \equiv \bot$: not $M \Vdash A$.
2. $A \equiv R(d_1, \dots, d_n)$: $M \Vdash A$ iff $\langle d_1^M, \dots, d_n^M \rangle \in R^M$.
3. $A \equiv d_1 = d_2$: $M \Vdash A$ iff $d_1^M = d_2^M$.
4. $A \equiv \neg B$: $M \Vdash A$ iff not $M \Vdash B$.
5. $A \equiv (B \wedge C)$: $M \Vdash A$ iff $M \Vdash B$ and $M \Vdash C$.
6. $A \equiv (B \vee C)$: $M \Vdash A$ iff $M \Vdash B$ or $M \Vdash C$ (or both).
7. $A \equiv (B \rightarrow C)$: $M \Vdash A$ iff not $M \Vdash B$ or $M \Vdash C$ (or both).
8. $A \equiv \forall x\, B$: $M \Vdash A$ iff for all $a \in |M|$, $M[a/c] \Vdash B[c/x]$, if c does not occur in B.
9. $A \equiv \exists x\, B$: $M \Vdash A$ iff there is an $a \in |M|$ such that $M[a/c] \Vdash B[c/x]$, if c does not occur in B.

Let $x_1, \dots, x_n$ be all free variables in A, $c_1, \dots, c_n$ constant symbols not in A, $a_1, \dots, a_n \in |M|$, and $s(x_i) = a_i$.

Show that $M, s \vDash A$ iff $M[a_1/c_1, \dots, a_n/c_n] \Vdash A[c_1/x_1] \dots [c_n/x_n]$.

(This problem shows that it is possible to give a semantics for first-order logic that makes do without variable assignments.)

Problem B.7. Suppose that f is a function symbol not in $A(x, y)$. Show that there is a structure M such that $M \vDash \forall x\, \exists y\, A(x, y)$ iff there is an M' such that $M' \vDash \forall x\, A(x, f(x))$.

(This problem is a special case of what's known as Skolem's Theorem; $\forall x\, A(x, f(x))$ is called a *Skolem normal form* of $\forall x\, \exists y\, A(x, y)$.)

Problem B.8. Carry out the proof of Proposition B.31 in detail.

Problem B.9. Prove Proposition B.34

Problem B.10.

1. Show that $\Gamma \vDash \bot$ iff Γ is unsatisfiable.
2. Show that $\Gamma \cup \{A\} \vDash \bot$ iff $\Gamma \vDash \neg A$.
3. Suppose c does not occur in A or Γ. Show that $\Gamma \vDash \forall x\, A$ iff $\Gamma \vDash A[c/x]$.

Problem B.11. Complete the proof of Proposition B.42.

APPENDIX C

Natural Deduction

C.1 Natural Deduction

Natural deduction is a derivation system intended to mirror actual reasoning (especially the kind of regimented reasoning employed by mathematicians). Actual reasoning proceeds by a number of "natural" patterns. For instance, proof by cases allows us to establish a conclusion on the basis of a disjunctive premise, by establishing that the conclusion follows from either of the disjuncts. Indirect proof allows us to establish a conclusion by showing that its negation leads to a contradiction. Conditional proof establishes a conditional claim "if ... then ..." by showing that the consequent follows from the antecedent. Natural deduction is a formalization of some of these natural inferences. Each of the logical connectives and quantifiers comes with two rules, an introduction and an elimination rule, and they each correspond to one such natural inference pattern. For instance, $\rightarrow$Intro corresponds to conditional proof, and $\lor$Elim to proof by cases. A particularly simple rule is $\land$Elim which allows the inference from $A \land B$ to A (or B).

One feature that distinguishes natural deduction from other derivation systems is its use of assumptions. A derivation in natural deduction is a tree of formulas. A single formula stands at the root of the tree of formulas, and the "leaves" of the tree are formulas from which the conclusion is derived. In natural deduction, some leaf formulas play a role inside the derivation but are "used up" by the time the derivation reaches the conclusion. This corresponds to the practice, in actual reasoning, of introducing hypotheses which only remain in effect for a short while. For instance, in a proof by cases, we assume the truth of each of the disjuncts; in conditional proof, we assume the truth of the antecedent; in indirect proof, we assume the truth of the negation of the conclusion. This way of introducing hypothetical assumptions and then doing away with them in the service of establishing an intermediate step is a hallmark of natural deduction. The formulas at the leaves of a natural deduction derivation are called assumptions, and some of the rules of inference may "discharge" them. For instance, if we have a derivation of B from some assumptions which include A, then the $\rightarrow$Intro rule allows us to infer $A \rightarrow B$ and discharge any assumption of the form A. (To keep track of which assumptions are discharged at which inferences, we label the inference and the assumptions it discharges with a number.) The assumptions that remain undischarged at the end of the derivation are together sufficient for the truth of the conclusion, and so a derivation establishes that its undischarged assumptions entail its conclusion.

The relation $\Gamma \vdash A$ based on natural deduction holds iff there is a derivation in which A is the last sentence in the tree, and every leaf which is undischarged is in Γ. A is a theorem in natural deduction iff there is a derivation in which A is the last sentence and all assumptions are discharged. For instance, here is a derivation that shows that $\vdash (A \wedge B) \rightarrow A$:

$$\cfrac{\cfrac{[A \wedge B]^1}{A}\ \wedge\text{Elim}}{(A \wedge B) \rightarrow A}\ {\rightarrow}\text{Intro}\ (1)$$

The label 1 indicates that the assumption $A \wedge B$ is discharged at the $\rightarrow$Intro inference.

A set Γ is inconsistent iff $\Gamma \vdash \bot$ in natural deduction. The rule $\bot_I$ makes it so that from an inconsistent set, any sentence can be derived.

Natural deduction systems were developed by Gerhard Gentzen and Stanisław Jaśkowski in the 1930s, and later developed by Dag Prawitz and Frederic Fitch. Because its inferences mirror natural methods of proof, it is favored by philosophers. The versions developed by Fitch are often used in introductory logic textbooks. In the philosophy of logic, the rules of natural deduction have sometimes been taken to give the meanings of the logical operators ("proof-theoretic semantics").

C.2 Rules and Derivations

Natural deduction systems are meant to closely parallel the informal reasoning used in mathematical proof (hence it is somewhat "natural"). Natural deduction proofs begin with assumptions. Inference rules are then applied. Assumptions are "discharged" by the $\neg$Intro, $\rightarrow$Intro, $\lor$Elim and $\exists$Elim inference rules, and the label of the discharged assumption is placed beside the inference for clarity.

Definition C.1 (Assumption). An *assumption* is any sentence in the topmost position of any branch.

Derivations in natural deduction are certain trees of sentences, where the topmost sentences are assumptions, and if a sentence stands below one, two, or three other sequents, it must follow correctly by a rule of inference. The sentences at the top of the inference are called the *premises* and the sentence below the *conclusion* of the inference. The rules come in pairs, an introduction and an elimination rule for each logical operator. They introduce a logical operator in the conclusion or remove a logical operator from a premise of the rule. Some of the rules allow an assumption of a certain type to be *discharged.* To indicate which assumption is discharged by which inference, we also assign labels to both the assumption and the inference. This is indicated by writing the assumption as "$[A]^n$."

It is customary to consider rules for all the logical operators $\wedge$, $\vee$, $\rightarrow$, $\neg$, and $\bot$, even if some of those are consider as defined.

C.3 Propositional Rules

Rules for $\wedge$

$$\frac{A \quad B}{A \wedge B}\ \wedge\text{Intro}$$

$$\frac{A \wedge B}{A}\ \wedge\text{Elim} \qquad \frac{A \wedge B}{B}\ \wedge\text{Elim}$$

Rules for $\vee$

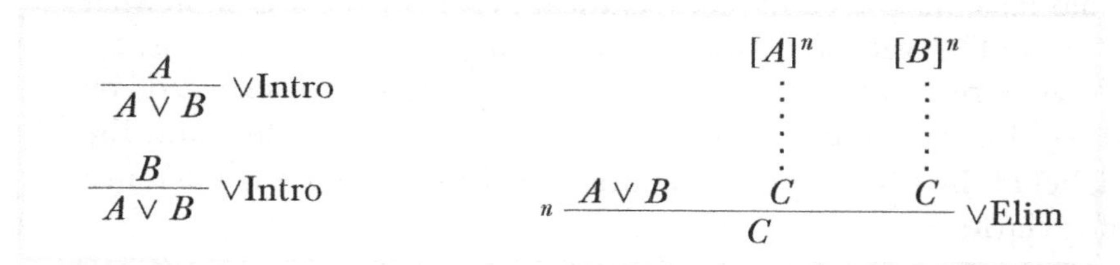

Rules for $\rightarrow$

$$\begin{array}{c}[A]^n \\ \vdots \\ B \end{array}$$
$$n\ \frac{B}{A \rightarrow B}\ \rightarrow\text{Intro}$$

$$\frac{A \rightarrow B \quad A}{B}\ \rightarrow\text{Elim}$$

Rules for $\neg$

Rules for ⊥

Note that ¬Intro and $\bot_C$ are very similar: The difference is that ¬Intro derives a negated sentence $\neg A$ but $\bot_C$ a positive sentence A.

Whenever a rule indicates that some assumption may be discharged, we take this to be a permission, but not a requirement. E.g., in the →Intro rule, we may discharge any number of assumptions of the form A in the derivation of the premise B, including zero.

C.4 Quantifier Rules

Rules for ∀

$$\frac{A(a)}{\forall x\, A(x)}\ \forall\text{Intro} \qquad\qquad \frac{\forall x\, A(x)}{A(t)}\ \forall\text{Elim}$$

In the rules for ∀, t is a ground term (a term that does not contain any variables), and a is a constant symbol which does not occur in the conclusion $\forall x\, A(x)$, or in any assumption which is undischarged in the derivation ending with the premise $A(a)$. We call a the *eigenvariable* of the ∀Intro inference.

Rules for $\exists$

$$\frac{A(t)}{\exists x\, A(x)}\ \exists\text{Intro}$$

$$\cfrac{\exists x\, A(x) \qquad \begin{array}{c}[A(a)]^n \\ \vdots \\ C\end{array}}{C}\ \exists\text{Elim}\ (n)$$

Again, t is a ground term, and a is a constant which does not occur in the premise $\exists x\, A(x)$, in the conclusion C, or any assumption which is undischarged in the derivations ending with the two premises (other than the assumptions $A(a)$). We call a the *eigenvariable* of the $\exists$Elim inference.

The condition that an eigenvariable neither occur in the premises nor in any assumption that is undischarged in the derivations leading to the premises for the $\forall$Intro or $\exists$Elim inference is called the *eigenvariable condition.*

We use the term "eigenvariable" even though a in the above rules is a constant. This has historical reasons.

In $\exists$Intro and $\forall$Elim there are no restrictions, and the term t can be anything, so we do not have to worry about any conditions. On the other hand, in the $\exists$Elim and $\forall$Intro rules, the eigenvariable condition requires that the constant symbol a does not occur anywhere in the conclusion or in an undischarged assumption. The condition is necessary to ensure that the system is sound, i.e., only derives sentences from undischarged assumptions from which they follow. Without this condition, the following would be allowed:

$$\cfrac{\exists x\, A(x) \qquad \cfrac{[A(a)]^1}{\forall x\, A(x)}\ {*}\forall\text{Intro}}{\forall x\, A(x)}\ \exists\text{Elim}$$

However, $\exists x\, A(x) \nvDash \forall x\, A(x)$.

C.5 Derivations

We've said what an assumption is, and we've given the rules of inference. Derivations in natural deduction are inductively generated from these: each derivation either is an assumption on its own, or consists of one, two, or three derivations followed by a correct inference.

Definition C.2 (Derivation). A *derivation* of a sentence A from assumptions Γ is a tree of sentences satisfying the following conditions:

1. The topmost sentences of the tree are either in Γ or are discharged by an inference in the tree.
2. The bottommost sentence of the tree is A.
3. Every sentence in the tree except the sentence A at the bottom is a premise of a correct application of an inference rule whose conclusion stands directly below that sentence in the tree.

We then say that A is the *conclusion* of the derivation and that A is *derivable* from Γ.

Example C.3. Every assumption on its own is a derivation. So, e.g., C by itself is a derivation, and so is D by itself. We can obtain a new derivation from these by applying, say, the $\wedge$Intro rule,

$$\frac{A \qquad B}{A \wedge B} \wedge\text{Intro}$$

These rules are meant to be general: we can replace the A and B in it with any sentences, e.g., by C and D. Then the conclusion would be $C \wedge D$, and so

$$\frac{C \qquad D}{C \wedge D} \wedge\text{Intro}$$

is a correct derivation. Of course, we can also switch the assumptions, so that D plays the role of A and C that of B. Thus,

$$\dfrac{D \qquad C}{D \wedge C} \wedge\text{Intro}$$

is also a correct derivation.

We can now apply another rule, say, →Intro, which allows us to conclude a conditional and allows us to discharge any assumption that is identical to the antecedent of that conditional. So both of the following would be correct derivations:

$$1 \dfrac{\dfrac{[C]^1 \qquad D}{C \wedge D} \wedge\text{Intro}}{C \rightarrow (C \wedge D)} \rightarrow\text{Intro} \qquad 1 \dfrac{\dfrac{C \qquad [D]^1}{C \wedge D} \wedge\text{Intro}}{D \rightarrow (C \wedge D)} \rightarrow\text{Intro}$$

Remember that discharging of assumptions is a permission, not a requirement: we don't have to discharge the assumptions. In particular, we can apply a rule even if the assumptions are not present in the derivation. For instance, the following is legal, even though there is no assumption A to be discharged:

$$1 \dfrac{B}{A \rightarrow B} \rightarrow\text{Intro}$$

C.6 Examples of Derivations

Example C.4. Let's give a derivation of the sentence $(A \wedge B) \rightarrow A$.

We begin by writing the desired conclusion at the bottom of the derivation.

$$\overline{(A \wedge B) \rightarrow A}$$

Next, we need to figure out what kind of inference could result in a sentence of this form. The main operator of the conclusion is →, so we'll try to arrive at the conclusion using the →Intro rule. It is best to write down the assumptions involved and label the inference rules as you progress, so it is easy to see whether all assumptions have been discharged at the end of the proof.

$$\dfrac{\begin{matrix}[A \wedge B]^1 \\ \vdots \\ A\end{matrix}}{(A \wedge B) \to A}\ {\to}\text{Intro}\ (1)$$

We now need to fill in the steps from the assumption $A \wedge B$ to A. Since we only have one connective to deal with, $\wedge$, we must use the $\wedge$ elim rule. This gives us the following proof:

$$1\ \dfrac{\dfrac{[A \wedge B]^1}{A}\ \wedge\text{Elim}}{(A \wedge B) \to A}\ {\to}\text{Intro}$$

We now have a correct derivation of $(A \wedge B) \to A$.

Example C.5. Now let's give a derivation of $(\neg A \vee B) \to (A \to B)$.

We begin by writing the desired conclusion at the bottom of the derivation.

$$\overline{(\neg A \vee B) \to (A \to B)}$$

To find a logical rule that could give us this conclusion, we look at the logical connectives in the conclusion: $\neg$, $\vee$, and $\to$. We only care at the moment about the first occurence of $\to$ because it is the main operator of the sentence in the end-sequent, while $\neg$, $\vee$ and the second occurence of $\to$ are inside the scope of another connective, so we will take care of those later. We therefore start with the $\to$Intro rule. A correct application must look like this:

$$1\ \dfrac{\begin{matrix}[\neg A \vee B]^1 \\ \vdots \\ A \to B\end{matrix}}{(\neg A \vee B) \to (A \to B)}\ {\to}\text{Intro}$$

This leaves us with two possibilities to continue. Either we can keep working from the bottom up and look for another application of the $\to$Intro rule, or we can work from the top down and apply a $\lor$Elim rule. Let us apply the latter. We will use the assumption $\neg A \lor B$ as the leftmost premise of $\lor$Elim. For a valid application of $\lor$Elim, the other two premises must be identical to the conclusion $A \to B$, but each may be derived in turn from another assumption, namely the two disjuncts of $\neg A \lor B$. So our derivation will look like this:

$$
\cfrac[l]{\cfrac{[\neg A \lor B]^1 \qquad \begin{matrix}[\neg A]^2 \\ \vdots \\ A \to B\end{matrix} \qquad \begin{matrix}[B]^2 \\ \vdots \\ A \to B\end{matrix}}{A \to B}\ {\scriptstyle 2}\ \lor\text{Elim}}{(\neg A \lor B) \to (A \to B)}\ {\scriptstyle 1}\ \to\text{Intro}
$$

In each of the two branches on the right, we want to derive $A \to B$, which is best done using $\to$Intro.

$$
\cfrac{\cfrac{[\neg A \lor B]^1 \qquad \cfrac{\begin{matrix}[\neg A]^2, [A]^3 \\ \vdots \\ B\end{matrix}}{A \to B}\ {\scriptstyle 3}\ \to\text{Intro} \qquad \cfrac{\begin{matrix}[B]^2, [A]^4 \\ \vdots \\ B\end{matrix}}{A \to B}\ {\scriptstyle 4}\ \to\text{Intro}}{A \to B}\ {\scriptstyle 2}\ \lor\text{Elim}}{(\neg A \lor B) \to (A \to B)}\ {\scriptstyle 1}\ \to\text{Intro}
$$

For the two missing parts of the derivation, we need derivations of B from $\neg A$ and A in the middle, and from A and B on the left. Let's take the former first. $\neg A$ and A are the two premises of $\neg$Elim:

$$
\begin{matrix}\cfrac{[\neg A]^2 \qquad [A]^3}{\bot}\ \neg\text{Elim} \\ \vdots \\ B\end{matrix}
$$

By using $\bot_I$, we can obtain B as a conclusion and complete the branch.

$$
\begin{prooftree}
\AxiomC{$[\neg A \lor B]^1$}
\AxiomC{$[\neg A]^2$}
\AxiomC{$[A]^3$}
\RightLabel{$\bot$Intro}
\BinaryInfC{$\bot$}
\RightLabel{$\bot_I$}
\UnaryInfC{B}
\LeftLabel{3}
\RightLabel{$\to$Intro}
\UnaryInfC{$A \to B$}
\AxiomC{$[B]^2, [A]^4$}
\noLine
\UnaryInfC{$\vdots$}
\noLine
\UnaryInfC{B}
\LeftLabel{4}
\RightLabel{$\to$Intro}
\UnaryInfC{$A \to B$}
\LeftLabel{2}
\RightLabel{$\lor$Elim}
\TrinaryInfC{$A \to B$}
\LeftLabel{1}
\RightLabel{$\to$Intro}
\UnaryInfC{$(\neg A \lor B) \to (A \to B)$}
\end{prooftree}
$$

Let's now look at the rightmost branch. Here it's important to realize that the definition of derivation ***allows assumptions to be discharged*** but ***does not require*** them to be. In other words, if we can derive B from one of the assumptions A and B without using the other, that's ok. And to derive B from B is trivial: B by itself is such a derivation, and no inferences are needed. So we can simply delete the assumption A.

$$
\begin{prooftree}
\AxiomC{$[\neg A \lor B]^1$}
\AxiomC{$[\neg A]^2$}
\AxiomC{$[A]^3$}
\RightLabel{$\neg$Elim}
\BinaryInfC{$\bot$}
\RightLabel{$\bot_I$}
\UnaryInfC{B}
\LeftLabel{3}
\RightLabel{$\to$Intro}
\UnaryInfC{$A \to B$}
\AxiomC{$[B]^2$}
\RightLabel{$\to$Intro}
\UnaryInfC{$A \to B$}
\LeftLabel{2}
\RightLabel{$\lor$Elim}
\TrinaryInfC{$A \to B$}
\LeftLabel{1}
\RightLabel{$\to$Intro}
\UnaryInfC{$(\neg A \lor B) \to (A \to B)$}
\end{prooftree}
$$

Note that in the finished derivation, the rightmost $\to$Intro inference does not actually discharge any assumptions.

Example C.6. So far we have not needed the $\bot_C$ rule. It is special in that it allows us to discharge an assumption that isn't a sub-formula of the conclusion of the rule. It is closely related to the $\bot_I$ rule. In fact, the $\bot_I$ rule is a special case of the $\bot_C$ rule—there is a logic called "intuitionistic logic" in which only $\bot_I$ is allowed. The $\bot_C$ rule is a last resort when nothing else works. For instance, suppose we want to derive $A \lor \neg A$. Our usual strategy would be to attempt to derive $A \lor \neg A$ using $\lor$Intro. But this would require us to derive either A or $\neg A$ from no assumptions, and this can't be done. $\bot_C$ to the rescue!

$$1\;\dfrac{\begin{array}{c}[\neg(A \vee \neg A)]^1\\ \vdots\\ \bot\end{array}}{A \vee \neg A}\;\bot_C$$

Now we're looking for a derivation of $\bot$ from $\neg(A \vee \neg A)$. Since $\bot$ is the conclusion of $\neg$Elim we might try that:

$$1\;\dfrac{\dfrac{\begin{array}{c}[\neg(A \vee \neg A)]^1\\ \vdots\\ \neg A\end{array}\qquad\begin{array}{c}[\neg(A \vee \neg A)]^1\\ \vdots\\ A\end{array}}{\bot}\;\neg\text{Elim}}{A \vee \neg A}\;\bot_C$$

Our strategy for finding a derivation of $\neg A$ calls for an application of $\neg$Intro:

$$1\;\dfrac{\dfrac{2\;\dfrac{\begin{array}{c}[\neg(A \vee \neg A)]^1, [A]^2\\ \vdots\\ \bot\end{array}}{\neg A}\;\neg\text{Intro}\qquad\begin{array}{c}[\neg(A \vee \neg A)]^1\\ \vdots\\ A\end{array}}{\bot}\;\neg\text{Elim}}{A \vee \neg A}\;\bot_C$$

Here, we can get $\bot$ easily by applying $\neg$Elim to the assumption $\neg(A \vee \neg A)$ and $A \vee \neg A$ which follows from our new assumption A by $\vee$Intro:

$$1\;\dfrac{\dfrac{2\;\dfrac{\dfrac{[\neg(A \vee \neg A)]^1\qquad \dfrac{[A]^2}{A \vee \neg A}\;\vee\text{Intro}}{\bot}\;\neg\text{Elim}}{\neg A}\;\neg\text{Intro}\qquad\begin{array}{c}[\neg(A \vee \neg A)]^1\\ \vdots\\ A\end{array}}{\bot}\;\neg\text{Elim}}{A \vee \neg A}\;\bot_C$$

On the right side we use the same strategy, except we get A by $\bot_C$:

$$
\dfrac{\dfrac{\dfrac{\dfrac{}{\dfrac{}{}\;} [\neg(A \lor \neg A)]^1 \quad \dfrac{[A]^2}{A \lor \neg A}\,\lor\text{Intro}}{\bot}\,\neg\text{Elim}}{\neg A}\,{}^{2}\,\neg\text{Intro} \qquad \dfrac{\dfrac{[\neg(A \lor \neg A)]^1 \quad \dfrac{[\neg A]^3}{A \lor \neg A}\,\lor\text{Intro}}{\bot}\,\neg\text{Elim}}{A}\,{}^{3}\,\bot_C}{\dfrac{\bot}{A \lor \neg A}\,{}^{1}\,\bot_C}\,\neg\text{Elim}
$$

C.7 Derivations with Quantifiers

Example C.7. When dealing with quantifiers, we have to make sure not to violate the eigenvariable condition, and sometimes this requires us to play around with the order of carrying out certain inferences. In general, it helps to try and take care of rules subject to the eigenvariable condition first (they will be lower down in the finished proof).

Let's see how we'd give a derivation of the formula $\exists x \neg A(x) \rightarrow \neg \forall x A(x)$. Starting as usual, we write

$$
\dfrac{}{\exists x \neg A(x) \rightarrow \neg \forall x A(x)}
$$

We start by writing down what it would take to justify that last step using the $\rightarrow$Intro rule.

$$
\dfrac{\begin{matrix}[\exists x \neg A(x)]^1 \\ \vdots \\ \neg \forall x A(x)\end{matrix}}{\exists x \neg A(x) \rightarrow \neg \forall x A(x)}\,{}^{1}\,\rightarrow\text{Intro}
$$

Since there is no obvious rule to apply to $\neg \forall x A(x)$, we will proceed by setting up the derivation so we can use the $\exists$Elim rule. Here we must pay attention to the eigenvariable condition, and choose a constant that does not appear in $\exists x A(x)$ or any assumptions that it depends on. (Since no constant symbols appear, however, any choice will do fine.)

$$
{\scriptstyle 1}\,\dfrac{{\scriptstyle 2}\,\dfrac{[\exists x\, \neg A(x)]^1 \qquad \begin{array}{c}[\neg A(a)]^2\\ \vdots\\ \neg \forall x\, A(x)\end{array}}{\neg \forall x\, A(x)}\ \exists\text{Elim}}{\exists x\, \neg A(x) \rightarrow \neg \forall x\, A(x)}\ {\rightarrow}\text{Intro}
$$

In order to derive $\neg \forall x\, A(x)$, we will attempt to use the $\neg$Intro rule: this requires that we derive a contradiction, possibly using $\forall x\, A(x)$ as an additional assumption. Of course, this contradiction may involve the assumption $\neg A(a)$ which will be discharged by the $\rightarrow$Intro inference. We can set it up as follows:

$$
{\scriptstyle 1}\,\dfrac{{\scriptstyle 2}\,\dfrac{[\exists x\, \neg A(x)]^1 \qquad {\scriptstyle 3}\,\dfrac{\begin{array}{c}[\neg A(a)]^2, [\forall x\, A(x)]^3\\ \vdots\\ \bot\end{array}}{\neg \forall x\, A(x)}\ \neg\text{Intro}}{\neg \forall x\, A(x)}\ \exists\text{Elim}}{\exists x\, \neg A(x) \rightarrow \neg \forall x\, A(x)}\ {\rightarrow}\text{Intro}
$$

It looks like we are close to getting a contradiction. The easiest rule to apply is the $\forall$Elim, which has no eigenvariable conditions. Since we can use any term we want to replace the universally quantified x, it makes the most sense to continue using a so we can reach a contradiction.

$$
{\scriptstyle 1}\,\dfrac{{\scriptstyle 2}\,\dfrac{[\exists x\, \neg A(x)]^1 \qquad {\scriptstyle 3}\,\dfrac{\dfrac{[\neg A(a)]^2 \qquad \dfrac{[\forall x\, A(x)]^3}{A(a)}\ \forall\text{Elim}}{\bot}\ \neg\text{Elim}}{\neg \forall x\, A(x)}\ \neg\text{Intro}}{\neg \forall x\, A(x)}\ \exists\text{Elim}}{\exists x\, \neg A(x) \rightarrow \neg \forall x\, A(x)}\ {\rightarrow}\text{Intro}
$$

It is important, especially when dealing with quantifiers, to double check at this point that the eigenvariable condition has not been violated. Since the only rule we applied that is subject to the eigenvariable condition was $\exists$Elim, and the eigenvariable a

does not occur in any assumptions it depends on, this is a correct derivation.

Example C.8. Sometimes we may derive a formula from other formulas. In these cases, we may have undischarged assumptions. It is important to keep track of our assumptions as well as the end goal.

Let's see how we'd give a derivation of the formula $\exists x\, C(x, b)$ from the assumptions $\exists x\, (A(x) \land B(x))$ and $\forall x\, (B(x) \to C(x, b))$. Starting as usual, we write the conclusion at the bottom.

$$\frac{}{\exists x\, C(x, b)}$$

We have two premises to work with. To use the first, i.e., try to find a derivation of $\exists x\, C(x, b)$ from $\exists x\, (A(x) \land B(x))$ we would use the $\exists$Elim rule. Since it has an eigenvariable condition, we will apply that rule first. We get the following:

$$1\,\frac{\exists x\, (A(x) \land B(x)) \qquad \begin{matrix}[A(a) \land B(a)]^1 \\ \vdots \\ \exists x\, C(x, b)\end{matrix}}{\exists x\, C(x, b)}\,\exists\text{Elim}$$

The two assumptions we are working with share B. It may be useful at this point to apply $\land$Elim to separate out $B(a)$.

$$1\,\frac{\exists x\, (A(x) \land B(x)) \qquad \begin{matrix}\dfrac{[A(a) \land B(a)]^1}{B(a)}\,\land\text{Elim} \\ \vdots \\ \exists x\, C(x, b)\end{matrix}}{\exists x\, C(x, b)}\,\exists\text{Elim}$$

The second assumption we have to work with is $\forall x\, (B(x) \to C(x, b))$. Since there is no eigenvariable condition we can instantiate x with the constant symbol a using $\forall$Elim to get $B(a) \to C(a, b)$.

We now have both $B(a) \to C(a,b)$ and $B(a)$. Our next move should be a straightforward application of the $\to$Elim rule.

$$\dfrac{\dfrac{\forall x\,(B(x) \to C(x,b))}{B(a) \to C(a,b)}\,\forall\text{Elim} \qquad \dfrac{[A(a) \wedge B(a)]^1}{B(a)}\,\wedge\text{Elim}}{C(a,b)}\,\to\text{Elim}$$

$$\begin{array}{c} \vdots \end{array}$$

$$1\ \dfrac{\exists x\,(A(x) \wedge B(x)) \qquad \exists x\, C(x,b)}{\exists x\, C(x,b)}\,\exists\text{Elim}$$

We are so close! One application of $\exists$Intro and we have reached our goal.

$$1\ \dfrac{\exists x\,(A(x) \wedge B(x)) \qquad \dfrac{\dfrac{\dfrac{\forall x\,(B(x) \to C(x,b))}{B(a) \to C(a,b)}\,\forall\text{Elim} \qquad \dfrac{[A(a) \wedge B(a)]^1}{B(a)}\,\wedge\text{Elim}}{C(a,b)}\,\to\text{Elim}}{\exists x\, C(x,b)}\,\exists\text{Intro}}{\exists x\, C(x,b)}\,\exists\text{Elim}$$

Since we ensured at each step that the eigenvariable conditions were not violated, we can be confident that this is a correct derivation.

Example C.9. Give a derivation of the formula $\neg\forall x\, A(x)$ from the assumptions $\forall x\, A(x) \to \exists y\, B(y)$ and $\neg\exists y\, B(y)$. Starting as usual, we write the target formula at the bottom.

$$\dfrac{}{\neg\forall x\, A(x)}$$

The last line of the derivation is a negation, so let's try using $\neg$Intro. This will require that we figure out how to derive a contradiction.

$$1\ \dfrac{\begin{array}{c}[\forall x\, A(x)]^1 \\ \vdots \\ \bot\end{array}}{\neg\forall x\, A(x)}\,\neg\text{Intro}$$

So far so good. We can use $\forall$Elim but it's not obvious if that will help us get to our goal. Instead, let's use one of our assumptions. $\forall x\, A(x) \to \exists y\, B(y)$ together with $\forall x\, A(x)$ will allow us to use the $\to$Elim rule.

$$\dfrac{\forall x\, A(x) \to \exists y\, B(y) \qquad [\forall x\, A(x)]^1}{\exists y\, B(y)}\ \to\text{Elim}$$

$$\vdots$$

$$1\ \dfrac{\bot}{\neg\forall x\, A(x)}\ \neg\text{Intro}$$

We now have one final assumption to work with, and it looks like this will help us reach a contradiction by using $\neg$Elim.

$$1\ \dfrac{\dfrac{\neg\exists y\, B(y) \qquad \dfrac{\forall x\, A(x) \to \exists y\, B(y) \qquad [\forall x\, A(x)]^1}{\exists y\, B(y)}\ \to\text{Elim}}{\bot}\ \neg\text{Elim}}{\neg\forall x\, A(x)}\ \neg\text{Intro}$$

C.8 Derivations with Identity predicate

Derivations with identity predicate require additional inference rules.

$$\dfrac{}{t = t}\ =\text{Intro} \qquad\qquad \dfrac{t_1 = t_2 \qquad A(t_1)}{A(t_2)}\ =\text{Elim} \qquad \dfrac{t_1 = t_2 \qquad A(t_2)}{A(t_1)}\ =\text{Elim}$$

In the above rules, t, t_1, and t_2 are closed terms. The =Intro rule allows us to derive any identity statement of the form $t = t$ outright, from no assumptions.

Example C.10. If s and t are closed terms, then $A(s), s = t \vdash A(t)$:

$$\frac{s = t \qquad A(s)}{A(t)}\ {=}\text{Elim}$$

This may be familiar as the "principle of substitutability of identicals," or Leibniz' Law.

Example C.11. We derive the sentence

$$\forall x\, \forall y\, ((A(x) \land A(y)) \rightarrow x = y)$$

from the sentence

$$\exists x\, \forall y\, (A(y) \rightarrow y = x)$$

We develop the derivation backwards:

$$\cfrac{\cfrac{1\ \cfrac{\begin{matrix}\exists x\, \forall y\, (A(y) \rightarrow y = x) \quad [A(a) \land A(b)]^1 \\ \vdots \\ a = b\end{matrix}}{((A(a) \land A(b)) \rightarrow a = b)}\ {\rightarrow}\text{Intro}}{\forall y\, ((A(a) \land A(y)) \rightarrow a = y)}\ \forall\text{Intro}}{\forall x\, \forall y\, ((A(x) \land A(y)) \rightarrow x = y)}\ \forall\text{Intro}$$

We'll now have to use the main assumption: since it is an existential formula, we use $\exists$Elim to derive the intermediary conclusion $a = b$.

$$\cfrac{\cfrac{1\ \cfrac{2\ \cfrac{\exists x\, \forall y\, (A(y) \rightarrow y = x) \qquad \begin{matrix}[\forall y\, (A(y) \rightarrow y = c)]^2 \\ [A(a) \land A(b)]^1 \\ \vdots \\ a = b\end{matrix}}{a = b}\ \exists\text{Elim}}{((A(a) \land A(b)) \rightarrow a = b)}\ {\rightarrow}\text{Intro}}{\forall y\, ((A(a) \land A(y)) \rightarrow a = y)}\ \forall\text{Intro}}{\forall x\, \forall y\, ((A(x) \land A(y)) \rightarrow x = y)}\ \forall\text{Intro}$$

The sub-derivation on the top right is completed by using its assumptions to show that $a = c$ and $b = c$. This requies two separate derivations. The derivation for $a = c$ is as follows:

$$\dfrac{\dfrac{[\forall y\,(A(y) \to y = c)]^2}{A(a) \to a = c}\;\forall\text{Elim} \qquad \dfrac{[A(a) \land A(b)]^1}{A(a)}\;\land\text{Elim}}{a = c}\;{\to}\text{Elim}$$

From $a = c$ and $b = c$ we derive $a = b$ by $=$Elim.

C.9 Proof-Theoretic Notions

Just as we've defined a number of important semantic notions (validity, entailment, satisfiabilty), we now define corresponding *proof-theoretic notions.* These are not defined by appeal to satisfaction of sentences in structures, but by appeal to the derivability or non-derivability of certain sentences from others. It was an important discovery that these notions coincide. That they do is the content of the *soundness* and *completeness theorems.*

Definition C.12 (Theorems). A sentence A is a *theorem* if there is a derivation of A in natural deduction in which all assumptions are discharged. We write $\vdash A$ if A is a theorem and $\nvdash A$ if it is not.

Definition C.13 (Derivability). A sentence A is *derivable from* a set of sentences Γ, $\Gamma \vdash A$, if there is a derivation with conclusion A and in which every assumption is either discharged or is in Γ. If A is not derivable from Γ we write $\Gamma \nvdash A$.

Definition C.14 (Consistency). A set of sentences Γ is *inconsistent* iff $\Gamma \vdash \bot$. If Γ is not inconsistent, i.e., if $\Gamma \nvdash \bot$, we say it is *consistent.*

Proposition C.15 (Reflexivity). *If $A \in \Gamma$, then $\Gamma \vdash A$.*

Proof. The assumption A by itself is a derivation of A where every undischarged assumption (i.e., A) is in Γ. □

Proposition C.16 (Monotony). *If $\Gamma \subseteq \Delta$ and $\Gamma \vdash A$, then $\Delta \vdash A$.*

Proof. Any derivation of A from Γ is also a derivation of A from Δ. □

Proposition C.17 (Transitivity). *If $\Gamma \vdash A$ and $\{A\} \cup \Delta \vdash B$, then $\Gamma \cup \Delta \vdash B$.*

Proof. If $\Gamma \vdash A$, there is a derivation δ_0 of A with all undischarged assumptions in Γ. If $\{A\} \cup \Delta \vdash B$, then there is a derivation δ_1 of B with all undischarged assumptions in $\{A\} \cup \Delta$. Now consider:

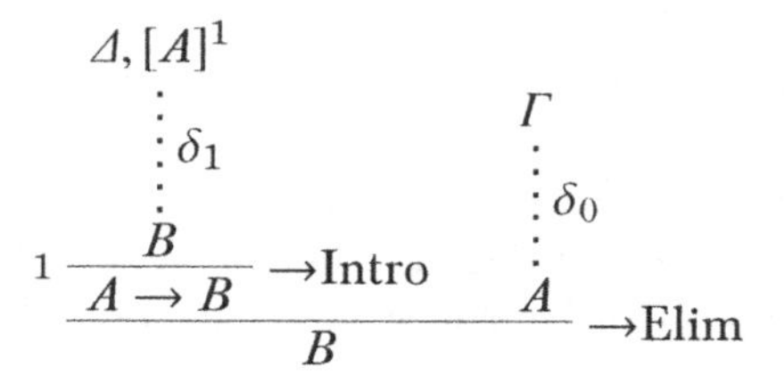

The undischarged assumptions are now all among $\Gamma \cup \Delta$, so this shows $\Gamma \cup \Delta \vdash B$. □

When $\Gamma = \{A_1, A_2, \ldots, A_k\}$ is a finite set we may use the simplified notation $A_1, A_2, \ldots, A_k \vdash B$ for $\Gamma \vdash B$, in particular $A \vdash B$ means that $\{A\} \vdash B$.

Note that if $\Gamma \vdash A$ and $A \vdash B$, then $\Gamma \vdash B$. It follows also that if $A_1, \ldots, A_n \vdash B$ and $\Gamma \vdash A_i$ for each i, then $\Gamma \vdash B$.

Proposition C.18. *Γ is inconsistent iff $\Gamma \vdash A$ for every sentence A.*

Proof. Exercise. □

Proposition C.19 (Compactness). *1. If $\Gamma \vdash A$ then there is a finite subset $\Gamma_0 \subseteq \Gamma$ such that $\Gamma_0 \vdash A$.*

2. If every finite subset of Γ is consistent, then Γ is consistent.

Proof. 1. If $\Gamma \vdash A$, then there is a derivation δ of A from Γ. Let Γ_0 be the set of undischarged assumptions of δ. Since any derivation is finite, Γ_0 can only contain finitely many sentences. So, δ is a derivation of A from a finite $\Gamma_0 \subseteq \Gamma$.

2. This is the contrapositive of (1) for the special case $A \equiv \bot$. □

Summary

Proof systems provide purely syntactic methods for characterizing consequence and compatibility between sentences. **Natural deduction** is one such proof system. A **derivation** in it consists of a tree of formulas. The topmost formula a derivation are **assumptions**. All other formulas, for the derivation to be correct, must be correctly justified by one of a number of **inference rules**. These come in pairs; an introduction and an elimination rule for each connective and quantifier. For instance, if a formula A is justified by a $\to$Elim rule, the preceding formulas (the **premises**) must be $B \to A$ and B (for some B). Some inference rules also allow assumptions to be **discharged**. For instance, if $A \to B$ is inferred from B using $\to$Intro, any occurrences of A as assumptions in the derivation leading to the premise B may be discharged, given a label that is also recorded at the inference.

If there is a derivation with end formula A and all assumptions are discharged, we say A is a theorem and write $\vdash A$. If all undischarged assumptions are in some set Γ, we say A is **derivable from** Γ and write $\Gamma \vdash A$. If $\Gamma \vdash \bot$ we say Γ is **inconsistent**, otherwise **consistent**. These notions are interrelated, e.g., $\Gamma \vdash A$ iff $\Gamma \cup \{\neg A\} \vdash \bot$. They are also related to the corresponding semantic notions, e.g., if $\Gamma \vdash A$ then $\Gamma \vDash A$. This property of natural deduction—what can be derived from Γ is guaranteed to be entailed by Γ—is called **soundness**. The **soundness theorem** is proved by induction on the length of derivations, showing that each individual inference preserves entailment of its conclusion from open assumptions provided its premises are entailed by their open assumptions.

Problems

Problem C.1. Give derivations of the following:

1. $\neg(A \rightarrow B) \rightarrow (A \land \neg B)$

2. $(A \rightarrow C) \lor (B \rightarrow C)$ from the assumption $(A \land B) \rightarrow C$

Problem C.2. Give derivations of the following:

1. $\exists y\, A(y) \rightarrow B$ from the assumption $\forall x\, (A(x) \rightarrow B)$

2. $\exists x\, (A(x) \rightarrow \forall y\, A(y))$

Problem C.3. Prove that $=$ is both symmetric and transitive, i.e., give derivations of $\forall x\, \forall y\, (x = y \rightarrow y = x)$ and $\forall x\, \forall y\, \forall z((x = y \land y = z) \rightarrow x = z)$

Problem C.4. Give derivations of the following formulas:

1. $\forall x\, \forall y\, ((x = y \land A(x)) \rightarrow A(y))$

2. $\exists x\, A(x) \land \forall y\, \forall z\, ((A(y) \land A(z)) \rightarrow y = z) \rightarrow \exists x\, (A(x) \land \forall y\, (A(y) \rightarrow y = x))$

Problem C.5. Prove Proposition C.18

APPENDIX D

Biographies

D.1 Alonzo Church

Fig. D.1: Alonzo Church

Alonzo Church was born in Washington, DC on June 14, 1903. In early childhood, an air gun incident left Church blind in one eye. He finished preparatory school in Connecticut in 1920 and began his university education at Princeton that same year. He completed his doctoral studies in 1927. After a couple years abroad, Church returned to Princeton. Church was known exceedingly polite and careful. His blackboard writing was immaculate, and he would preserve important papers by carefully covering them in Duco cement (a clear glue). Outside of his academic pursuits, he enjoyed reading science fiction magazines and was not afraid to write to the editors if he spotted any inaccuracies in the writing.

Church's academic achievements were great. Together with his students Stephen Kleene and Barkley Rosser, he developed a theory of effective calculability, the lambda calculus, independently of Alan Turing's development of the Turing machine. The two definitions of computability are equivalent, and give rise to what is now known as the *Church-Turing Thesis*, that a function of the natural numbers is effectively computable if and only if it is computable via Turing machine (or lambda calculus). He also proved what is now known as *Church's Theorem*: The decision problem for the validity of first-order formulas is unsolvable.

Church continued his work into old age. In 1967 he left Princeton for UCLA, where he was professor until his retirement in 1990. Church passed away on August 1, 1995 at the age of 92.

Further Reading For a brief biography of Church, see Enderton (2019). Church's original writings on the lambda calculus and the Entscheidungsproblem (Church's Thesis) are Church (1936a,b). Aspray (1984) records an interview with Church about the Princeton mathematics community in the 1930s. Church wrote a series of book reviews of the *Journal of Symbolic Logic* from 1936 until 1979. They are all archived on John MacFarlane's website (MacFarlane, 2015).

D.2 Kurt Gödel

Kurt Gödel (GER-dle) was born on April 28, 1906 in Brünn in the Austro-Hungarian empire (now Brno in the Czech Republic). Due to his inquisitive and bright nature, young Kurtele was often called "Der kleine Herr Warum" (Little Mr. Why) by his family. He excelled in academics from primary school onward, where he got less than the highest grade only in mathematics. Gödel was often absent from school due to poor health and was exempt from physical education. He was diagnosed with rheumatic fever during his childhood. Throughout his life, he believed this permanently affected his heart despite medical assessment saying otherwise.

Gödel began studying at the University of Vienna in 1924 and completed his doctoral studies in 1929. He first intended to study physics, but his interests soon moved to mathematics and especially logic, in part due to the influence of the philosopher Rudolf Carnap. His dissertation, written under the supervision of Hans Hahn, proved the completeness theorem of first-order predicate logic with identity (Gödel, 1929). Only a year later, he obtained his most famous results—the first and second incompleteness theorems (published in Gödel 1931). During his time in Vienna, Gödel was heavily involved with the Vienna Circle, a group of scientifically-minded philosophers that included Carnap, whose work was especially influenced by Gödel's results.

In 1938, Gödel married Adele Nimbursky. His parents were not pleased: not only was she six years older than him and already divorced, but she worked as a dancer in a nightclub. Social pressures did not affect Gödel, however, and they remained happily married until his death.

Fig. D.2: Kurt Gödel

After Nazi Germany annexed Austria in 1938, Gödel and Adele emigrated to the United States, where he took up a position at the Institute for Advanced Study in Princeton, New Jersey. Despite his introversion and eccentric nature, Gödel's time at Princeton was collaborative and fruitful. He published essays in set theory, philosophy and physics. Notably, he struck up a particularly strong friendship with his colleague at the IAS, Albert Einstein.

In his later years, Gödel's mental health deteriorated. His wife's hospitalization in 1977 meant she was no longer able to cook his meals for him. Having suffered from mental health issues throughout his life, he succumbed to paranoia. Deathly afraid of being poisoned, Gödel refused to eat. He died of starvation on January 14, 1978, in Princeton.

Further Reading For a complete biography of Gödel's life is available, see John Dawson (1997). For further biographical pieces, as well as essays about Gödel's contributions to logic and philosophy, see Wang (1990), Baaz et al. (2011), Takeuti et al. (2003), and Sigmund et al. (2007).

Gödel's PhD thesis is available in the original German (Gödel, 1929). The original text of the incompleteness theorems is (Gödel, 1931). All of Gödel's published and unpublished writings, as well as a selection of correspondence, are available in English in his *Collected Papers* Feferman et al. (1986, 1990).

For a detailed treatment of Gödel's incompleteness theorems, see Smith (2013). For an informal, philosophical discussion of Gödel's theorems, see Mark Linsenmayer's podcast (Linsenmayer, 2014).

D.3 Rózsa Péter

Rózsa Péter was born Rósza Politzer, in Budapest, Hungary, on February 17, 1905. She is best known for her work on recursive functions, which was essential for the creation of the field of recursion theory.

Péter was raised during harsh political times—WWI raged when she was a teenager—but was able to attend the affluent Maria Terezia Girls' School in Budapest, from where she graduated in 1922. She then studied at Pázmány Péter University (later renamed Loránd Eötvös University) in Budapest. She began studying chemistry at the insistence of her father, but later switched to mathematics, and graduated in 1927. Although she had the credentials to teach high school mathematics, the economic situation at the time was dire as the Great Depression affected the world economy. During this time, Péter took odd jobs as a tutor and private teacher of mathematics. She eventually returned to university to take up graduate studies in mathematics. She had originally planned to work in number theory, but after finding out that her results had already been proven, she almost gave up on mathematics altogether. She was encouraged to work on Gödel's incompleteness theorems, and unknowingly proved several of his results in different ways. This restored her confidence, and Péter went on to write her first papers on recursion theory, inspired by David Hilbert's foundational program. She received her PhD in 1935, and in 1937 she became an editor for the *Journal of Symbolic Logic.*

Péter's early papers are widely credited as founding contributions to the field of recursive function theory. In Péter (1935a), she investigated the relationship between different kinds of recursion. In Péter (1935b), she showed that a certain recursively defined function is not primitive recursive. This simplified an earlier result due to Wilhelm Ackermann. Péter's simplified function is what's now often called the Ackermann function—and sometimes, more properly, the Ackermann-Péter function. She wrote the first book on recursive function theory (Péter, 1951).

Fig. D.3: Rózsa Péter

Despite the importance and influence of her work, Péter did not obtain a full-time teaching position until 1945. During the Nazi occupation of Hungary during World War II, Péter was not allowed to teach due to anti-Semitic laws. In 1944 the government created a Jewish ghetto in Budapest; the ghetto was cut off from the rest of the city and attended by armed guards. Péter was forced to live in the ghetto until 1945 when it was liberated. She then went on to teach at the Budapest Teachers Training College, and from 1955 onward at Eötvös Loránd University. She was the first female Hungarian mathematician to become an Academic Doctor of Mathematics, and the first woman to be elected to the Hungarian Academy of Sciences.

Péter was known as a passionate teacher of mathematics, who preferred to explore the nature and beauty of mathematical problems with her students rather than to merely lecture. As a result, she was affectionately called "Aunt Rosa" by her students. Péter died in 1977 at the age of 71.

Further Reading For more biographical reading, see (O'Connor and Robertson, 2014) and (Andrásfai, 1986). Tamassy (1994) conducted a brief interview with Péter. For a fun read about mathematics, see Péter's book *Playing With Infinity* (Péter, 2010).

D.4 Julia Robinson

Julia Bowman Robinson was an American mathematician. She is known mainly for her work on decision problems, and most famously for her contributions to the solution of Hilbert's tenth problem. Robinson was born in St. Louis, Missouri on December 8, 1919. At a young age Robinson recalls being intrigued by numbers (Reid, 1986, 4). At age nine she contracted scarlet fever and suffered from several recurrent bouts of rheumatic fever. This forced her to spend much of her time in bed, putting her behind in her education. Although she was able to catch up with the help of private tutors, the physical effects of her illness had a lasting impact on her life.

Despite her childhood struggles, Robinson graduated high school with several awards in mathematics and the sciences. She started her university career at San Diego State College, and transferred to the University of California, Berkeley as a senior. There she was highly influenced by mathematician Raphael Robinson. They quickly became good friends, and married in 1941. As a spouse of a faculty member, Robinson was barred from teaching in the mathematics department at Berkeley. Although she continued to audit mathematics classes, she hoped to leave university and start a family. Not long after her wedding, however, Robinson contracted pneumonia. She was told that there was substantial scar tissue build up on her heart due to the rheumatic fever she suffered as a child. Due to the severity of the scar tissue, the doctor predicted that she would not live past forty and she was advised not to have children (Reid, 1986, 13).

Fig. D.4: Julia Robinson

Robinson was depressed for a long time, but eventually decided to continue studying mathematics. She returned to Berkeley and completed her PhD in 1948 under the supervision of Alfred Tarski. The first-order theory of the real numbers had been shown to be decidable by Tarski, and from Gödel's work it followed that the first-order theory of the natural numbers is undecidable. It was a major open problem whether the first-order theory of the rationals is decidable or not. In her thesis (1949), Robinson proved that it was not.

Interested in decision problems, Robinson next attempted to find a solution Hilbert's tenth problem. This problem was one of a famous list of 23 mathematical problems posed by David Hilbert in 1900. The tenth problem asks whether there is an algorithm that will answer, in a finite amount of time, whether or not a polynomial equation with integer coefficients, such as $3x^2 - 2y + 3 = 0$, has a solution in the integers. Such questions are known as *Diophantine problems.* After some initial successes, Robinson joined forces with Martin Davis and Hilary Putnam, who were also working on the problem. They succeeded in showing that exponential Diophantine problems (where the unknowns may also appear as exponents) are undecidable, and showed that a certain conjecture (later called "J.R.") implies that Hilbert's tenth problem is undecidable (Davis et al., 1961). Robinson continued to work on the problem for the next decade. In 1970, the young Russian mathematician Yuri Matijasevich finally proved the J.R. hypothesis. The combined result is now called the Matijasevich-Robinson-Davis-Putnam theorem, or MDRP theorem for short. Matijasevich and Robinson became friends and collaborated on several papers. In a letter to Matijasevich, Robinson once wrote that "actually I am very pleased that working together (thousands of miles apart) we are obviously making more progress than either one of us could alone" (Matijasevich, 1992, 45).

Robinson was the first female president of the American Mathematical Society, and the first woman to be elected to the National Academy of Science. She died on July 30, 1985 at the age of 65 after being diagnosed with leukemia.

Further Reading Robinson's mathematical papers are available in her *Collected Works* (Robinson, 1996), which also includes a reprint of her National Academy of Sciences biographical memoir (Feferman, 1994). Robinson's older sister Constance Reid published an "Autobiography of Julia," based on interviews (Reid, 1986), as well as a full memoir (Reid, 1996). A short documentary about Robinson and Hilbert's tenth problem was directed by George Csicsery (Csicsery, 2016). For a brief memoir about Yuri Matijasevich's collaborations with Robinson, and her influence on his work, see (Matijasevich, 1992).

D.5 Alfred Tarski

Fig. D.5: Alfred Tarski

Alfred Tarski was born on January 14, 1901 in Warsaw, Poland (then part of the Russian Empire). Described as "Napoleonic," Tarski was boisterous, talkative, and intense. His energy was often reflected in his lectures—he once set fire to a wastebasket while disposing of a cigarette during a lecture, and was forbidden from lecturing in that building again.

Tarski had a thirst for knowledge from a young age. Although later in life he would tell students that he studied logic because it was the only class in which he got a B, his high school records show that he got A's across the board—even in logic. He studied at the University of Warsaw from 1918 to 1924. Tarski first intended to study biology, but became interested in mathematics, philosophy, and logic, as the university was the center of the Warsaw School of Logic and Philosophy. Tarski earned his doctorate in 1924 under the supervision of Stanisław Leśniewski.

Before emigrating to the United States in 1939, Tarski completed some of his most important work while working as a secondary school teacher in Warsaw. His work on logical consequence and logical truth were written during this time. In 1939, Tarski was visiting the United States for a lecture tour. During his visit, Germany invaded Poland, and because of his Jewish heritage, Tarski could not return. His wife and children remained in Poland until the end of the war, but were then able to emigrate to the United States as well. Tarski taught at Harvard, the College of the City of New York, and the Institute for Advanced Study at Princeton, and finally the University of California, Berkeley.

There he founded the multidisciplinary program in Logic and the Methodology of Science. Tarski died on October 26, 1983 at the age of 82.

Further Reading For more on Tarski's life, see the biography *Alfred Tarski: Life and Logic* (Feferman and Feferman, 2004). Tarski's seminal works on logical consequence and truth are available in English in (Corcoran, 1983). All of Tarski's original works have been collected into a four volume series, (Tarski, 1981).

Photo Credits

Alonzo Church, p. 253: Portrait of Alonzo Church, undated, photographer unknown. Alonzo Church Papers; 1924–1995, (C0948) Box 60, Folder 3. Manuscripts Division, Department of Rare Books and Special Collections, Princeton University Library. © Princeton University. The Open Logic Project has obtained permission to use this image for inclusion in non-commercial OLP-derived materials. Permission from Princeton University is required for any other use.

Kurt Gödel, p. 255: Portrait of Kurt Gödel, ca. 1925, photographer unknown. From the Shelby White and Leon Levy Archives Center, Institute for Advanced Study, Princeton, NJ, USA, on deposit at Princeton University Library, Manuscript Division, Department of Rare Books and Special Collections, Kurt Gödel Papers, (C0282), Box 14b, #110000. The Open Logic Project has obtained permission from the Institute's Archives Center to use this image for inclusion in non-commercial OLP-derived materials. Permission from the Archives Center is required for any other use.

Rózsa Péter, p. 257: Portrait of Rózsa Péter, undated, photographer unknown. Courtesy of Béla Andrásfai.

Julia Robinson, p. 259: Portrait of Julia Robinson, unknown photographer, courtesy of Neil D. Reid. The Open Logic Project has obtained permission to use this image for inclusion in non-commercial OLP-derived materials. Permission is required for any other use.

Alfred Tarski, p. 261: Passport photo of Alfred Tarski, 1939. Cropped and restored from a scan of Tarski's passport by Joel Fuller. Original courtesy of Bancroft Library, University of California, Berkeley. Alfred Tarski Papers, Banc MSS 84/49. The Open Logic Project has obtained permission to use this image for inclusion in non-commercial OLP-derived materials. Permission from Bancroft Library is required for any other use.

Bibliography

Andrásfai, Béla. 1986. Rózsa (Rosa) Péter. *Periodica Polytechnica Electrical Engineering* 30(2-3): 139–145. URL http://www.pp.bme.hu/ee/article/view/4651.

Aspray, William. 1984. The Princeton mathematics community in the 1930s: Alonzo Church. URL http://www.princeton.edu/mudd/finding_aids/mathoral/pmc05.htm. Interview.

Baaz, Matthias, Christos H. Papadimitriou, Hilary W. Putnam, Dana S. Scott, and Charles L. Harper Jr. 2011. *Kurt Gödel and the Foundations of Mathematics: Horizons of Truth.* Cambridge: Cambridge University Press.

Church, Alonzo. 1936a. A note on the Entscheidungsproblem. *Journal of Symbolic Logic* 1: 40–41.

Church, Alonzo. 1936b. An unsolvable problem of elementary number theory. *American Journal of Mathematics* 58: 345–363.

Corcoran, John. 1983. *Logic, Semantics, Metamathematics.* Indianapolis: Hackett, 2nd ed.

Csicsery, George. 2016. Zala films: Julia Robinson and Hilbert's tenth problem. URL http://www.zalafilms.com/films/juliarobinson.html.

Davis, Martin, Hilary Putnam, and Julia Robinson. 1961. The decision problem for exponential Diophantine equations. *Annals of Mathematics* 74(3): 425–436. URL http://www.jstor.org/stable/1970289.

Enderton, Herbert B. 2019. Alonzo Church: Life and Work. In *The Collected Works of Alonzo Church*, eds. Tyler Burge and Herbert B. Enderton. Cambridge, MA: MIT Press.

Feferman, Anita and Solomon Feferman. 2004. *Alfred Tarski: Life and Logic.* Cambridge: Cambridge University Press.

Feferman, Solomon. 1994. Julia Bowman Robinson 1919–1985. *Biographical Memoirs of the National Academy of Sciences* 63: 1–28. URL http://www.nasonline.org/publications/biographical-memoirs/memoir-pdfs/robinson-julia.pdf.

Feferman, Solomon, John W. Dawson Jr., Stephen C. Kleene, Gregory H. Moore, Robert M. Solovay, and Jean van Heijenoort. 1986. *Kurt Gödel: Collected Works. Vol. 1: Publications 1929–1936.* Oxford: Oxford University Press.

Feferman, Solomon, John W. Dawson Jr., Stephen C. Kleene, Gregory H. Moore, Robert M. Solovay, and Jean van Heijenoort. 1990. *Kurt Gödel: Collected Works. Vol. 2: Publications 1938–1974.* Oxford: Oxford University Press.

Gödel, Kurt. 1929. Über die Vollständigkeit des Logikkalküls [On the completeness of the calculus of logic]. Dissertation, Universität Wien. Reprinted and translated in Feferman et al. (1986), pp. 60–101.

Gödel, Kurt. 1931. über formal unentscheidbare Sätze der *Principia Mathematica* und verwandter Systeme I [On formally undecidable propositions of *Principia Mathematica* and related systems I]. *Monatshefte für Mathematik und Physik* 38: 173–198. Reprinted and translated in Feferman et al. (1986), pp. 144–195.

John Dawson, Jr. 1997. *Logical Dilemmas: The Life and Work of Kurt Gödel.* Boca Raton: CRC Press.

Linsenmayer, Mark. 2014. The partially examined life: Gödel on math. URL http://www.partiallyexaminedlife.com/2014/06/16/ep95-godel/. Podcast audio.

MacFarlane, John. 2015. Alonzo Church's JSL reviews. URL http://johnmacfarlane.net/church.html.

Matijasevich, Yuri. 1992. My collaboration with Julia Robinson. *The Mathematical Intelligencer* 14(4): 38–45.

O'Connor, John J. and Edmund F. Robertson. 2014. Rózsa Péter. URL http://www-groups.dcs.st-and.ac.uk/~history/Biographies/Peter.html.

Péter, Rózsa. 1935a. Über den Zusammenhang der verschiedenen Begriffe der rekursiven Funktion. *Mathematische Annalen* 110: 612–632.

Péter, Rózsa. 1935b. Konstruktion nichtrekursiver Funktionen. *Mathematische Annalen* 111: 42–60.

Péter, Rózsa. 1951. *Rekursive Funktionen.* Budapest: Akademiai Kiado. English translation in (Péter, 1967).

Péter, Rózsa. 1967. *Recursive Functions.* New York: Academic Press.

Péter, Rózsa. 2010. *Playing with Infinity.* New York: Dover. URL https://books.google.ca/books?id=6V3wNs4uv_4C&lpg=PP1&ots=BkQZaHcR99&lr&pg=PP1#v=onepage&q&f=false.

Reid, Constance. 1986. The autobiography of Julia Robinson. *The College Mathematics Journal* 17: 3–21.

Reid, Constance. 1996. *Julia: A Life in Mathematics.* Cambridge: Cambridge University Press. URL https://books.google.ca/books?id=lRtSzQyHf9UC&lpg=PP1&pg=PP1#v=onepage&q&f=false.

Robinson, Julia. 1949. Definability and decision problems in arithmetic. *Journal of Symbolic Logic* 14(2): 98–114. URL http://www.jstor.org/stable/2266510.

Robinson, Julia. 1996. *The Collected Works of Julia Robinson*. Providence: American Mathematical Society.

Sigmund, Karl, John Dawson, Kurt Mühlberger, Hans Magnus Enzensberger, and Juliette Kennedy. 2007. Kurt Gödel: Das Album–The Album. *The Mathematical Intelligencer* 29(3): 73–76.

Smith, Peter. 2013. *An Introduction to Gödel's Theorems*. Cambridge: Cambridge University Press.

Takeuti, Gaisi, Nicholas Passell, and Mariko Yasugi. 2003. *Memoirs of a Proof Theorist: Gödel and Other Logicians*. Singapore: World Scientific.

Tamassy, Istvan. 1994. Interview with Róza Péter. *Modern Logic* 4(3): 277–280.

Tarski, Alfred. 1981. *The Collected Works of Alfred Tarski*, vol. I–IV. Basel: Birkhäuser.

Wang, Hao. 1990. *Reflections on Kurt Gödel*. Cambridge: MIT Press.

About the Open Logic Project

The *Open Logic Text* is an open-source, collaborative textbook of formal meta-logic and formal methods, starting at an intermediate level (i.e., after an introductory formal logic course). Though aimed at a non-mathematical audience (in particular, students of philosophy and computer science), it is rigorous.

Coverage of some topics currently included may not yet be complete, and many sections still require substantial revision. We plan to expand the text to cover more topics in the future. We also plan to add features to the text, such as a glossary, a list of further reading, historical notes, pictures, better explanations, sections explaining the relevance of results to philosophy, computer science, and mathematics, and more problems and examples. If you find an error, or have a suggestion, please let the project team know.

The project operates in the spirit of open source. Not only is the text freely available, we provide the LaTeX source under the Creative Commons Attribution license, which gives anyone the right to download, use, modify, re-arrange, convert, and re-distribute our work, as long as they give appropriate credit. Please see the Open Logic Project website at openlogicproject.org for additional information.

Made in the USA
Monee, IL
23 October 2023